Science and Literature in Cormac McCarthy's Expanding Worlds

Science and Literature in Cormac McCarthy's Expanding Worlds

Bryan Giemza

BLOOMSBURY ACADEMIC
NEW YORK · LONDON · OXFORD · NEW DELHI · SYDNEY

BLOOMSBURY ACADEMIC
Bloomsbury Publishing Inc
1385 Broadway, New York, NY 10018, USA
50 Bedford Square, London, WC1B 3DP, UK
29 Earlsfort Terrace, Dublin 2, Ireland

BLOOMSBURY, BLOOMSBURY ACADEMIC and the Diana logo are trademarks
of Bloomsbury Publishing Plc

First published in the United States of America 2023

For legal purposes the List of Figures and Acknowledgments on p. 161 constitute
an extension of this copyright page.

Cover design: Eleanor Rose
Cover image: Cormac Macarthy photographed in Antartica, February 13 2010
© Todd Murphy

Bloomsbury Publishing Inc does not have any control over, or responsibility for, any
third-party websites referred to or in this book. All internet addresses given in this
book were correct at the time of going to press. The author and publisher regret any
inconvenience caused if addresses have changed or sites have ceased to exist,
but can accept no responsibility for any such changes.

Library of Congress Cataloging-in-Publication Data

Names: Giemza, Bryan Albin, author.
Title: Science and literature in Cormac McCarthy's expanding worlds / Bryan Giemza.
Description: New York: Bloomsbury Academic, 2023. | Includes bibliographical references. |
Summary: "The first book on Cormac McCarthy's engagement with the natural sciences,
paving the way for discussions on both McCarthy's collected works to date and the
intersections of the humanities and science"– Provided by publisher.
Identifiers: LCCN 2022049393 (print) | LCCN 2022049394 (ebook) | ISBN 9781501383779
(hardback) | ISBN 9781501383816 (paperback) | ISBN 9781501383786 (ebook) |
ISBN 9781501383793 (pdf) | ISBN 9781501383809 (ebook other)
Subjects: LCSH: McCarthy, Cormac, 1933–Criticism and interpretation. |
McCarthy, Cormac, 1933–Themes, motives. | Science in literature.
Classification: LCC PS3563.C337 Z663 2023 (print) | LCC PS3563.C337 (ebook) |
DDC 813/.54–dc23/eng/20230131

LC record available at https://lccn.loc.gov/2022049393

LC ebook record available at https://lccn.loc.gov/2022049394

ISBN: HB: 978-1-5013-8377-9
ePDF: 978-1-5013-8379-3
eBook: 978-1-5013-8378-6

Typeset by Deanta Global Publishing Services, Chennai, India

To find out more about our authors and books visit www.bloomsbury.com and
sign up for our newsletters.

To my daughter, Vera Rose, who sees deeply.
In gratitude for seeing with her.

CONTENTS

FIGURES

1

Introduction

The Trail to Santa Fe and the Stars (and Why It's Good *Craic*)

There isn't any place like the Santa Fe Institute, and there isn't any writer like Cormac, so the two fit quite well together.

—MURRAY GELL-MANN[1]

Perhaps the principal thing I absorbed from Professor Prall, and from Harvard in general, was a sense of interdisciplinary values— that the best way to "know" a thing is in the context of another discipline.

—LEONARD BERNSTEIN, *THE UNANSWERED QUESTION* (1976)[2]

The main campus of the Santa Fe Institute (SFI) (the Institute) is a lofty place, relative to the town center. At the top of a foothill in the Sangre de Cristo range, it sits a squat architectural cypher in the ever-changeable, pellucid light-scapes and shadow-scapes of New Mexico's high desert. Current president David Krakauer likes to say that the Institute "liv[es] at the edge of wilderness and society."[3] Enter the renovated core of what was once a manse's main office and you will see where the padrone once sat at a desk, blindingly backlit by sun-facing windows, throwing the dazzled, floodlit visitor into sharp illumination, and the head of the house into a place of shadowy observation. Home to a handful of resident faculty and a rotating cast of external and visiting scholars, the place receives its share of

FIGURE 1.1 *Santa Fe sunset from the ridge of Cormac McCarthy's never-finished former residence, March 2021, photographed by the author.*

nonhuman visitors, too, mutely attested by the coyote tracks crisscrossing the grounds.

The subject of SFI's inquiry, broadly, is complex systems science—not exactly a household term. To try to elicit a glimmer of recognition when describing it, turn to the 1990s, when complexity science wandered briefly into pop culture salience. Although the critical mass of scientific unities formed around complexity in the 1980s, it took another decade, sped by an explosion of accessible computer modeling and visualization, to creep into more public visibility with the popularity of Stephen Hawking, chaos theory and fractals, and Douglas Hofstadter's *Gödel, Escher, Bach: An Eternal Golden Braid* (known to insiders simply as GEB).[4] When the uninitiated try to pin him down about what SFI does, Krakauer's first line is to crib SFI co-founding member Murray Gell-Mann, the quark-theory pioneer (and, as Cormac points out, its likely co-discoverer).[5] What they do, he says, is "take a crude look at the whole," a phrase that SFI's first postdoctoral fellow, economist John H. Miller, gives to the title of his book offering a primer on complexity.[6]

More specifically, "SFI scientists seek the shared patterns and regularities across physical, biological, social, and technological systems that give rise to complexity—in any system in which its collective, system-wide behaviors cannot be understood merely by studying its parts or individuals in isolation." And, as Miller's book argues, "Insights from complexity science are increasingly useful in understanding questions far beyond the boundaries of traditional academic disciplines—urban sustainability, disease networks, and financial risk, to name a few."[7] Lately SFI has moved into avant-garde queries embracing the astrobiological—*what are the prospects for the evolution of life on other planets?* The guiding questions on the Institute's web page sliders today show how it has moved well beyond the popular culture of GEB days and the 1990s pop culture appeal of the butterfly effect: *Can algorithms bend toward justice? How does complexity apply to emergent political systems, musics, and pandemics?* Not exactly household questions either, taken as further evidence that SFI, like the symbols and equations that adorn its walls, nooks, and crannies (it is designed so that equations and questions can be scrawled throughout—on chalkboards and office windows alike), is a place for modern-day alchemists and arcana.

This reputation is reinforced by the Greek inscription at the door, apocryphally attributed to Plato's academy, translated as "Let no one ignorant of geometry enter." In my interpretation, it is to say if you have no passion for abstraction or the deep structures of mathematics as a language for understanding profound truths about the universe, well, you are knocking at the wrong door. Fittingly, a *Newsweek* reporter once anointed Murray Gell-Mann as Aristotle to SFI's Lyceum.[8] The comparison is more honest than forbidding, but the literary wonk in me still swooned a little before the gates of the academy on a hill.

Adding to the sensation of a place apart, around campus and on the spines of SFI Press imprints I spotted the alien characters of a glyphic alphabet designed by Brian Williams, as of this writing the creative director of Google Global Creative Works. Described as a means of "graphic enhancement and a system of categorization in SFI's book publishing and collateral," the glyphs are a familiar signifier of SFI's presence and events, if somewhat cryptic to outside eyes, in an institution that by its nature attracts actual cryptologists and language programmers.[9] It put me in mind of the Cherokee syllabary attributed to Sequoyah. My research in that area was a schooling in the complexity of writing systems and led me to debunk the reputedly divine origin of that alphabet (heavenly inspiration is something of a motif in the received histories of many writing systems, where in this case it appears that proselytizing missionaries, one of whom was trained specifically in orthography, ginned it up). Unsurprisingly, writing systems and linguistics are other areas of interest for SFI, as their newsfeed attests ("Study: Complexity holds steady as writing systems evolve").[10] And perhaps it points to the underlying desire to be the locus where new ways of inscripting otherworldly knowledge happen.

SFI wears the whispery side of its reputation comfortably and as an affirmation of curiosity's value, a reputation undoubtedly enhanced by its longtime de facto writer in residence, Cormac McCarthy, whose notoriety for remoteness has also been at times both carefully managed and cheerfully untended. Like any place where visits are mostly invitational, and whose founders and faculty are sometimes involved in opaque scientific work largely inscrutable to nonspecialists, imagination and rumor have colorized its popular reputation, styling the place as a mysterious enclave, reachable, as Flannery O'Connor used to say of her exotically named farm Andalusia, only by bus or buzzard.

This is somewhat overdone and yet apposite. Though waves of development and wildfires are fracturing its discreet landscapes, Santa Fe is currently home to just 85,000 souls, and often feels smaller, its built structures mostly thoughtfully and unobtrusively integrated into the landscape. SFI sits toward the edge of city limits, its gates plainly and prominently marked. Something about the human mind sees height as sovereignty. In claiming its burg in the Sangre de Cristo foothills, SFI enjoys the prominence of other well-known exalted places such as Museum Hill, St. John's College (a partner to the Institute dating back to its unhoused days), and the toney Ten Thousand Waves Spa, each, in its way, a center for different forms of cultural worship, in one of those rare American locales enamored with an aesthetic of its own making.

The Santa Fe Institute calls itself as "metaphorically a Monastery in the Mountains"—emphasis on the careful and ancient scholastic metaphor, in which one thing can be another.[11] When it started in 1983, SFI amounted to a circle of roaming intellectual compadres, mostly at Los Alamos, who hatched an organization with the rudiments of a post office box and a residential landline, holding their meetings in borrowed spaces and offices. Speaking on behalf of zoning accommodations for SFI in 1993, Doyne Farmer (now external SFI faculty and the founder of Prediction Company) was moved to say, "in 50 years, Santa Fe will be famous not for the Santa Fe Opera or for Georgia O'Keeffe but the Santa Fe Institute."[12] In August 1994 SFI held the first open house for its current main campus.[13] According to the 2020 edition of the *Global Go To Think Tank Index Report*, it now holds a place within the world's twenty-five ranked "Top Science and Technology Think Tanks" and "Best Transdisciplinary Research Think Tanks."[14]

And it is growing. In 2019, SFI cut the ribbon on its renovation of a second Miller Campus in the valleys of Tesuque, five miles from the perch of the original Cowan Campus. The Miller Campus continues the metaphor, "helping to bring the ideas from the 'Monastery' back to the 'Metropolis,'" as the SFI webpages would have it, supporting "the introduction and application of complexity science's big ideas and insights into the world."[15] Tesuque is derived from the Hispanicized Tewa place name *'Tat' unge' onwi* meaning "cottonwood place." The Tesuque Creek watershed and its bottom landscape are comparatively lush, calling to human visitors and inhabitants for over 5,000 years, and thus a suitable place for the "metropolitan" branch of SFI.[16] Even so, its existence is something of a miracle, the large and rare parcel of land a legacy from the 2012 gift of Eugene and Clare Thaw.[17] As a tight-knit community of creatives at a remove from suburban come-latelies, Tesuque likes to keep it that way, so procuring the place, the permissions, and (*marveille!*) the needful parking required both diplomacy and covenants. Here is the village at the foot of the hill, worldly but hardly ordinary, and as mundane piedmont dwellers have always appreciated, a livable and practical place.

The blessing of enduring institutions is that they are beholden to their founding vision and traditions; the curse of institutions is that they are forever yoked to narratives of their traditions, in the way that grown children inherit the typecasting of early family years. How the young Miller sibling campus's identity will mature remains to be seen. Complementarities can catalyze growth, and a combination of several visionary principles informing SFI's origin story might be partly credited for its steady expansion. The place was an early arrival to the interdisciplinary paradigm, now *de rigeur* in academe and think tanks. As I perused strings of newspapers articles from the 1980s about SFI, a stock phrase, amusingly quaint, arose about its insouciant disciplinary perspective: *this is not how things are done in higher research.*

Even some of those who brainstormed it into existence might have agreed. For example, Stephen Wolfram's interest was piqued when he was hailed to an October 1984 workshop by a solicitation from Nicholas Metropolis, the Greek-American, Oppenheimer-recruited Manhattan project scientist. Wolfram traveled from Princeton's Institute for Advanced Studies to the first enclave that would define the structure of what was then called the Rio Grande Institute, and only later, SFI. In Santa Fe, he found himself in "a slightly dark room, decorated with Native American artifacts. Around it were tables arranged in a large rectangle, at which sat a couple dozen men (yes, all men), mostly in their sixties," ostensibly to discuss the shape of "a new kind of teaching and research institute" dedicated, on a new model, to complexity science.[18]

Soon Wolfram, a London-born particle physicist who turned toward computer science, wearied of the pie-in-the-sky academic discussion. At twenty-five, Wolfram was the wunderkind at the table, having finished his PhD, received a MacArthur Fellowship, and joined the Caltech faculty at the tender age of twenty-one. Now he could play the enfant terrible. In his recollection, in the dusky boardroom of Santa Fe's School for Advanced Research,

> it seemed to have fallen to me to play the "let's get real" role. (To be fair, I had founded my first tech company a couple of years earlier, and wasn't a complete stranger to the world of grandiose "what-if" discussions, even if I was surprised, though more than a little charmed, to be seeing them in the sixty-something-year-old set.) A fragment of my notes from the day record [*sic*] my feelings: *What is supposed to be the point of this discussion?*[19]

Those familiar with higher education will recognize that what was happening at the table, actually, was sub-specialist jockeying for a piece of the Institute action, notwithstanding the mission to avoid academic fiefdoms. And when Wolfram gave voice to his concerns, he says he sensed in George Cowan, the meeting leader, "a mixture of frustration and relief at my question." (Wolfram played the part of a meeting *difficultator*, close kin to the facilitator, and just as needful.) Gell-Mann lobbied for an institute of complexity *and* simplicity, and Wolfram would argue that the Institute "should focus on what [he] called 'Complex Systems Theory.'"[20]

While this was explicit in the call of the meeting, in Wolfram's telling, he made bold to point to an actual *science* of complexity, and to call it that, inspired partly by his own work with cellular automata in which simple programming could produce very complex results. Like the senior most anglophile English professors who snarked my field when I was in graduate school ("I'll study American literature when there *is* one," said one), there are still those who dispute the validity of Wolfram's work, and indeed, whether complexity theory yields a "science" so much as a mash-up of scientific knowledge and mathematical approaches, which matters very little to Wolfram. SFI seems to be having the last word. "By the early 1990s," Wolfram writes, "probably in no small part through the efforts of the Santa Fe Institute, 'complexity' had actually become a popular buzzword, and, partly through a rather circuitous connection to climate science, funding had started pouring in."[21] SFI finds its stock buoyed once again by the relevance of climate change, AI and neural networks, and the complex systems that have undermined democracy globally, and, grimly, the hastening of the Doomsday Clock and rising awareness of the Institute's archetypal founding fears of extinction events and mass destruction.

SFI was ahead of its time in other respects, too. Presaging the later formulas of dot-com workplaces, the Institute insisted on common spaces and a rotation of visitors to avoid settling into stagnation. As its website states, "SFI's founders dreamt of an 'institute without walls,' where scientists could pursue pressing problems without regard to traditional disciplines or funding streams."[22] When I visited in May 2021, I found it an institute without doors, too. Visiting scholars work in open offices to encourage constant collaboration. In a scientific version of calling the question, anyone can summon the whole cohort to wrestle with a thorny problem when the occasion arises. "We always court a high risk of failure," Cormac McCarthy writes of the place. "Above all we have more fun than should be legal."[23]

Why would Cormac McCarthy say that scientists are more interesting conversationalists, as a class, than the tribe of fiction writers and authors to which he ostensibly belongs? He has explained affection for the Santa Fe Institute not just in terms of multidimensional, free-range critical thinking, but with reference to "craic," an Irish term with an etymology that goes to "conversation." That hardly covers it, though. When the Irish say, "What's the craic?" they are asking, *What's the story, and is it lively?* The term, loaded with cultural freight, is difficult to define outside of the culture. It can be applied to any stimulating session—including the artistically creative and aesthetically critical—that creates spiritual and intellectual communion between people. Above all, though, it entails a sense of fun, wonder, and exchange. No wonder McCarthy uses it to explain why he chose Santa Fe Institute (and it chose him): "But I'm here because I like science, and this is a fun place to spend time. There's good craic."[24]

Pursuing scientific critical approaches to McCarthy makes for good craic in the classroom, too. I teach mostly aspiring scientists, pre-med students, and engineers, and have shamelessly packaged my senior seminar with an eye-catching, pop culture title ("America's Great Novelist and the Zombie Apocalypse"), even though there is little truth in advertising, and students eventually learn that McCarthy's zombies take the form of allegorical *Road*-rats. Many of my students have not read a complete novel since high school, and they find themselves arrested, transfixed, and maddened by his work. Some perceive his work to be cheerless and hard, bludgeoning them into reluctant submission to his lesson that "ruder forms survive." Yet I can count, almost to the beat, on that moment when a biology major spots the evolutionary subtext of the phrase and gets pulled into the entwining of devolution and evolution in *The Road*. In examining latter day apocalypticism, we look at modern survivalist culture, play a values-clarifying bomb shelter "game," and see it dramatized in the film *After the Dark* (2013). Those activities tend to lead to discussions of Kant and utilitarianism, and sometimes to unexpected places, such as the deep biological signatures of historical cannibalism encoded through prion disease immunity, and whether Kant's moral imperative could be used to set moral bounds, or at least bend toward justice, within the parameters of artificial intelligence.

In fact, as a matter of teaching praxis, why not use McCarthy's works to flip the literary model and run with science altogether? We still read our share of literary criticism, but in meeting students where they are, my fictional horizons have been expanded in measure, and McCarthy's scientific vestments have become more fully legible to all of us. My students have created illuminated Google Earth voyages, dropping in from space to visit the landscapes of *Blood Meridian*, the better to understand how geography and natural history really do shape "destiny," manifest and otherwise. They have applied text-mining to identify patterns of use in McCarthy's works, looking at his wordstock to see if indeed its archaic texture owes something to a preference for Anglo-Saxon roots or later Romance Latinisms. They have risen to the challenge of assignments in narrative theory, empathy, and cognition, going on to create narrative-driven video games with complex outcomes. They have embraced archival research in Texas Tech's Sowell Family Collection in Literature, Community, and the Natural World, where they peruse the papers of scientists, ecologists, and various creatives, including some of McCarthy's correspondents such as Barry Lopez and Edward Abbey (see Chapter 4), to better understand McCarthy's wide reading and sources. Their assignments are necessarily synthetic, requiring the application and reconciliation of scientific and humanistic critical literature.

Convergently, in honoring the spirit of McCarthy's work, and the mutual reinforcement of scientific and humanistic discourses, our inquiries come to resemble the stuff of Santa Fe's big questions, and in framing those queries carefully, we become, in the broadest sense, a band of merry scientists. Recall the Institute's mission focus: scientists (n.b.) pursue pressing problems at SFI, along with occasional science-minded visitors and humanists. "If you know more than anybody else about a subject we want to talk to you. We don't care what the subject is," writes McCarthy, though the way to true interdisciplinarity, while refreshingly broad, remains, in the usual way, somewhat fuzzy—what is science but a way of knowing, whose practitioners form a motley and fruitfully fractured guild?[25] Like most institutions, the Institute is prudently wary of mission creep, and guarded about reputation. Complex systems science remains first among equals, or at least the polestar, in a remarkably wide ranging, raucous, and searching conversation.

Welcome to the craic in the conversation. Riffing on the notion of an exploded universe, this volume places McCarthy's work within contemporary scientific discourse and literary criticism. It challenges the myth of the solitary genius, in both scientific and humanistic endeavors. As a critical project, across these pages I hope to climb, like Jack's beanstalk or the Jacob's ladder of DNA, the twining strands of a STEM/humanities rope—up from the caverns of *The Road*, and along the fuse leading to the skies above Knoxville, blown to smithereens in *Suttree*. Chapter by chapter, the book probes, with respect to McCarthy's fictional universe and life experiences, the STEM subjects: science (chirality and the relationship between handedness and dis/order), technology (via the history of dynamite in east Tennessee), and engineering and the built environment (the Tennessee Valley Authority's larger-than-life, *hypanthropic* influence). "Hypanthropic" is a term archeologist Christopher Witmore uses to describe our current supra-anthropocene era in which human engineering takes place on a fully geological scale, "approaching that which is in excess of monstrosity" (see Chapter 4). As to math, the final chapter considers McCarthy's friendship with Guy Davenport and their shared interest in creating a unified aesthetic theory, culminating in McCarthy's essays on the origins of science, language, and the subconscious, and his final literary two-volume diptych, *The Passenger* and *Stella Maris* (2022), books centered on a mentally ill, mathematical genius and her brother.

Yes, science writ large (including social sciences and science-derived fields) holds the controlling interest of the SFI enterprise. I make no such claims here, since, as a humanist, I live in a discourse that is often carefree about launching its investigations from unfalsifiable hypotheses or using theory to create a (circular) logic of the falsifiable for the sake of its own inquiries. This book, which is mostly interested in contextual and translational work between humanities and conventionally understood STEM disciplines, has no theory of everything to offer readers, just case-specific insights from seeing McCarthy's work translationally.

Some of the articles that essay this territory[26] have resulted in (charitably) beautiful failures, betraying a deficient understanding, or the over-broad application, of scientific theory.[27] Indeed, Jean Bricmont, physicist, philosopher, and coauthor of *Fashionable Nonsense: Postmodern Intellectuals' Abuse of Science* (1999), skewered this sort of thing decades ago, writing that "Some purported 'applications' of chaos theory—for example, to business management or literary analysis—border on the absurd."[28] Applying scientific theory to humanities in a less than rigorous way weakens the general value proposition, reducing the incentive to build interdisciplinary communities—in the classroom and across university departments—that bond together for the communal discovery of knowledge, SFI-style.

It does a serious disservice to McCarthy's work as well. For example, it obscures the way *No Country for Old Men*'s coin toss addresses a counterintuitive concept, popularly misunderstood, that chaos is actually a deterministic phenomenon (chaos theory, like the complexity theory as the core of SFI, is intrinsically interdisciplinary).[29] It cuts off a nuanced discussion of how Bohmian mechanics posits an alternative formulation of quantum mechanics in opposition to the Copenhagen and Many Worlds interpretations, topics carefully parsed in McCarthy's 2022 duology of novels, *The Passenger* and *Stella Maris* (henceforth referred to as simply the duology). These distinctions need not be evasive to humanists or to literary narrative. And without them, we miss not only the full richness of what McCarthy's works have to offer, we lose out on the sheer joy of discovering the music between the lines of narrative and scientific theory.

Several important entries on complexity and McCarthy have come into view since the writing of this book began, redirecting it in turn: Laurence Gonzales's proposed history of the Santa Fe Institute, Ciarán Dowd's eye-opening chapter on the Santa Fe Institute in Steven Frye's *Cormac McCarthy in Context* (2020), and Lydia Cooper's *Cormac McCarthy: A Complexity Theory of Literature* (2021).[30] A senior technical writer for Google, Dowd offers an insider's perspective in his nuanced overview of SFI's influence on McCarthy, and vice versa. In addition to his chapter, until Cooper's book, no single scholarly monograph had addressed the subject. Cooper is certainly one of the best literary critics at work today among McCarthy scholars. She knows his works intimately and is expert at drawing connections between them to deliver larger insights, with an especially fine-tuned sense of their ethical dimensions. And she excels at bringing together disparate commentators and writing about them eloquently. In her book about complexity and Cormac McCarthy, she canvasses topics such as Thomas Piketty's capital versus wages problem and brings readers closer to the heart of a complex, contemporary economic dilemma. Accordingly, Dianne Luce's back cover blurb hails Cooper's book for its "confident perception of the overarching values that unify McCarthy's body of work."

After reading *A Complexity Theory of Literature*, however, one might wonder whether a complexity theory of literature successfully surfaces a set of overarching values in McCarthy's work. The complexity theory of literature, it turns out, is the notion that narrative helps us to make sense of complex systems because "narrative is a complex system." So, for example, the nexus between McCarthy's writings on capitalism and a "complexity theory of literature" is ultimately that narrative engages with economy, which is a complex system. And while economics are a main point of focus in the study, so it goes with related "ecosystems, forms of production, legislation, national identity, and psychology." The next step in the syllogism is to submit that McCarthy's SFI residency expanded his interest in evolutionary economics "and in complexity theory more broadly."[31] Given. And so? Cooper discovers other applications with occasional resort to the structuralist analysis that characterizes much of her earlier work, creating a converse form/function formula: simplistic semiotics indicate complex notions, and so on.

Cooper shares the understandable reluctance of complexity theorists to define complexity. As with any science, it brims with subfields and major areas of inquiry, and here a picture is worth a thousand words: understanding how emergent systems become self-organized, whether at the level of cellular biology, the construction of artificial intelligence, or strategic acting across societies, requires a concert of disciplines.

Cormac McCarthy himself accepts this ambiguity: "I don't think you can ask any ten people in science what complexity is and get a consistent view. I'm not going to go into it. The Santa Fe Institute basically deals with complexity in different disciplines. And there is a common thread to it, but it's kind of hard to come up with something that would satisfy everybody."[32] Humanists (including myself) show a sportive willingness to tour other fields with a kind of portmanteau intellectualism, bringing back souvenirs (or, um, in lit-crit parlance, *lenses*, like cheap sunglasses) to "apply," whereas the beauty of any purportedly scientific inquiry is the relentless cross-examination arising from experimentation and observation. These latter intellectual endeavors are free-standing in a different way. If we take the proposition that "narrative (and especially McCarthy's narrative) helps us to make sense of complex systems," what's the null hypothesis, and what will it take to make the case? In Cooper's framing, there is no experiment to be conducted or alternative hypothesis, only a case to be made by observation. It is not difficult to make a case for the general, but the work lies in the specific.

This is not Cooper's problem per se but one of humanities and its increasingly professionalized conventions, sometimes conceived in the spirit of science-envy, leading down a trackless path to the irrelevance that drives endless Defense of Humanities essays, delivered from the proscenium, while a career extinction event plays out behind the curtains. Moreover, in the arid

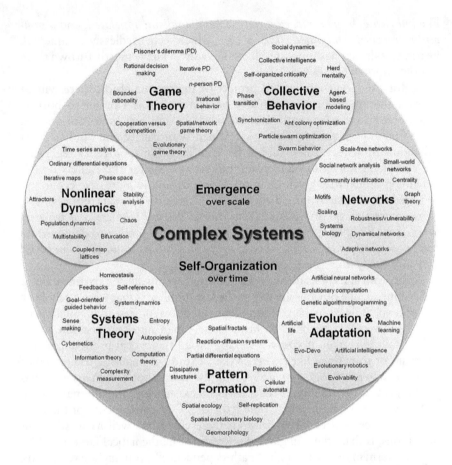

FIGURE 1.2 *This schematic usefully illustrates how encompassing Complex Systems are, and some of the subfields and major areas of Complex Systems Science. Illustration by Hiroki Sayama, DSc, Collective Dynamics of Complex Systems (CoCo) Research Group at Binghamton University, State University of New York. Reproduced here by ShareAlike 3.0 Unported license.*

field of literary-critical labor, where only certain conclusions are allowed, the ascendancy of grievance studies has led to formulaic critiques which in many cases are old wines served in a new skin. If we conceive of this in complexity terms, scholarship is a fractal enterprise that creates illusions of new forms (and complexity) from the same initial simplicities. The traces of certain well-trod paths are based on the idea that literary inquiry demands a pseudoscientific "critical methodology," so as to produce "verifiable" results. Well, even if they are not verifiable, they are certainly replicable. Those consigned to the critical salt mines and publishing conventions of contemporary English departments are compelled to signify with terms like *imbricate, hegemonic, excoriating neoliberalism, appelated,* and *a non-anthropocentric embodied ethics of care,* which seem needlessly ornamental, bearing a salt that has long since lost its savor. So I have mostly thrown them out the window.

As for the larger intellectual project of the present study, there will be little of literary theory or scholarly namechecking as I prefer to borrow a credo from the editors of *Nautilus* who "believe any subject in science, no matter how complex, can be explained with clarity and vitality."[33] At this point in history, there should be little need for apologetics regarding the complementarity of humanities and science (or for the endless op-eds defending humanities given their value proposition in a creative economy) or to redress the partly false dichotomy of C. P. Snow's "Two Cultures." The more interesting questions are: *How should literary scholars and teachers bring these disciplines rigorously and meaningfully together to demonstrate their necessary unities? How should they teach others to do the same? And why don't they do so more often?*

Apart from ancient academic traditions that eventually compartmentalized these subjects, one obvious answer to the last problem is that from a scholarly perspective, it is extremely demanding, in terms of research, intellectual labor, and received training, to bring these topics into conversation.[34] The risk of failure is high. If you are seeking broad theoretical pronouncements to result, or something like a poetics (i.e., a study of linguistic technique) of complex systems theory, you are reading the wrong book. Perhaps a poetics is emergent in the totality; if so, it owes its existence to the alternate idiom of STEM that permeates the whole structure. One consequence of the *pret-a-porter* conceptual chapters between these covers, as well as word count limitations, is that where unglossed scientific or mathematical jargon trickles in (say, a term in this chapter like "path-dependence"), you might have to look it up, as I did, with the reassurance that these are rewarding excursions. For example, in *Stella Maris,* Alicia confesses her disappointment in mathematics in the early pages: "In this case [the rebellion] was led by a group of evil and aberrant and wholly malicious partial differential equations who had conspired to usurp their own reality from the questionable circuitry of its creator's brain not unlike the rebellion which Milton describes and to fly their colors as an independent nation unaccountable to God or man alike" (10).

To understand what she is referring to, it helps to know *Paradise Lost,* and more, to have either a very knowledgeable friend or a deep understanding of the history of mathematics.[35] As a topologist, Alicia is gesturing not just to the mutiny of numbers, but to the Ricci flow, the tool eventually used by Grigori Perelman to prove the Poincaré conjecture. Of course, McCarthy offers no notes for any part of his math/science duology, just a series of knowing references and inside jokes, and so, unless you happen to be part of the SFI audience/readership, a great deal of the pleasure in reading it comes from the fact that the reader must take on the required side reading and other homework for comprehension. These latest books—which delayed the publication of this one, since it seemed strange to publish on the topic of science and math in McCarthy's world without at least glimpsing the motherlode—are likely to stand as McCarthy's *Finnegan's Wake*, useful to literary history and for keeping scholars busy for quite some time.

But there is no use in pretending those 600+ pages can be glossed in a chapter. They became available to me in the late stages of writing this book. Had they arrived sooner, they might have furnished the material for this entire study. As it stands, I will take a small step in that direction in the final chapter, eagerly anticipating the electricity the books will bring to a broader critical literary reconsideration of McCarthy's scientific turn of mind. The duology offers a moveable feast of complexity science through a century of mathematical and scientific dialogue, and McCarthy offers characteristically incisive readings of major developments in each, albeit in a wrapper that will be impenetrable to many readers. There are volumes to be written to explicate them, so to start with we might think about what they teach us about how they are to be read.

It's no accident that McCarthy's searching mind drew him to SFI, and that he perceived it as a center of real intellectual action, avoiding literary salons in like degree. In response to critics of his "The Kekulé Problem" essay—which concerns August Kekulé's dream of a snake that delivered, in a flash, the configuration of the benzene molecule that he had been trying to work out—McCarthy turned to Nietzsche and the argument that all human knowledge is simply our way of telling our own stories.[36] Narrative theorists and scientists have formed strange alliances in their understanding of the arrow of time, something that even scientists cannot describe without resort to narrative. Perhaps McCarthy needed complexity theory because literature didn't. But here we have arrived at another moonglow conversation: can narrative merely illustrate complexity theory, or complexity theory merely illustrate narrative? The answer is a transdisciplinary *neither*, *both*, *and*, and much of the translational work remains to be done someplace beyond the reward systems of humanities departments as they are currently constituted.

Or perhaps it's already happening in the halls of the Santa Fe Institute. One thoughtful reader of my project suggested that it should produce a poetics of complexity theory for McCarthy's work. While the suggestion is warmly appreciated, it seems to me oxymoronic. Even if it could be done, it would have the net effect of claiming scientific concepts for humanistic narrative via the idea of poetics. And even if it is not oxymoronic, it would be redundant, insofar as McCarthy's *The Passenger/Stella Maris* duology can already be understood as a poetics of math and science, and a purist rendering at that. Anything I might say about it will be surpassed by the original artifact. It takes a McCarthy to write a poetics of STEM, and he seems to have done just that with his newly published duology.

Moreover, in a sense, extrapolating such a poetics in literary criticism would be similar to describing quantum phenomena purely in terms of Newtonian physics, and it would put an unnecessary veneer of the literary on what is more accurately said through scientific poetics. It was precisely this dilemma that illuminated McCarthy's path to SFI in the first place, as he despaired of boozy, high-flung conversations with artsy scenesters who were articulating, in a different and imprecise argot, philosophies probed more rigorously in the language of mathematics, by scientists. Not the last word, but perhaps the best, and perhaps withal another human enterprise that should be informed by a tenor of radical humility since its outer edges have a way of folding in and collapsing (see Chapter 5). Cormac McCarthy wrote in a 1984 letter to Guy Davenport that reading (perhaps an advance copy of) *Surely You're Joking, Mr. Feynman!* was "the most fun I've had reading in a year,"[37] so I will borrow a Feynmanism, bequeathed to him by his first wife: what do *you* care what other people think? What I mean is that poetics would be the wrong mode for the book I set out to write and its peculiar methodology in pursuit of curious subjects, a way that was suggested by years of conversations with colleagues and what I noticed from the melding of creative subjectivities.

This may come across as duplicitous since I side with those who would puncture the myth of the solitary intellectual conquistador. A set of rediscovered early interviews with Cormac during his early and seemingly more voluble stages of career challenges the often-gendered idea of the writer-as-moody-recluse; taken with his later interviews, including with Oprah Winfrey, they show him as an inveterate conversationalist as well as a shrewd manager of his own reputation when it suits him.[38] SFI's principles reject the notion of solitary discovery, and so do Warren Bennis and Patricia Ward in *Organizing Genius: The Secrets of Creative Collaboration* (2007). Inspired by a conversation with Margaret Mead about "how networks of gifted people have changed the world," more so than talented people working alone, Biederman arrived at a theory of Great Groups (Mead called them "Sapiential Circles"). Bennis points to the Manhattan Project, of which SFI is a brainchild, as the "paradigmatic great group," but offers many other examples, including the original Walt Disney studio, Apple, the Clinton Campaign of 1992, Lockheed's Skunk Works, and the aesthetic school at Black Mountain. Bennis tugs at the common threads of great groups: they often start out in shabby places (like garages and office buildings), they spur each other on (thus avoiding blind alley defeatism), and they are most often "fueled by an invigorating, completely unrealistic view of what they can accomplish."[39]

Anyone well acquainted with SFI's history and culture is likely to grin as they check one box after another on Bennis's list of Great Groups criteria, including SFI's admirable devotion to the unrealistic. In fact, SFI started out with some loose talk of an endowment in the billions. In addition to institutional origination stories, finances have a way of revealing values in their true dimensions, as well as turns of fate that temper and sharpen vision. Six of eight of SFI's founding scientists worked at the Los Alamos labs. The Nobel Laureate of the cohort, Murray Gell-Mann, a physicist from Caltech, brought prospective fundraising starpower:

> Gell-Mann wanted what he termed "three units," by which he meant $300 million in endowments. A 1984 vision statement enthuses: "Within a few years of its inception, the Institute will require an endowment of approximately two hundred and fifty million dollars." (SFI's modern-day budget is approximately $10 million.)[40]

"I take on 50 times as much as the normal person, and I work at something like two percent efficiency," Gell-Mann said in an interview ten years later with the *Santa Fe New Mexican*.[41] For a $10M annual operating budget based solely upon the usual endowment draw-down formulas, SFI would need around two "units" to subsist on endowment alone. Tax records for the 501(c)(3) from the past decade confirm annual operating expenses from $10 million to $12 million. For the latest available tax year (2020), SFI holds net assets of $47,016,842, with a net endowment of $14,444,017 (compare this to Princeton's Institute for Advanced Studies, which Richard Feynman criticized, currently endowed at around $785M).[42] Among think tanks, then, SFI arguably punches well above its financial weight.

Between the dream and the reality, or, in Cormackian terms, "the wish and the thing" fell the shadow, which, with time, shifted. In SFI's two-installment website history, which as of this writing ends in the 1980s, David Pines recalls that "By the end of 1987 we were in the [Cristo Rey] convent, we had funding from NSF, funding from DOE, and funding from Citicorp. We were a healthy toddler about to enter nursery school."[43] It would take another six years of fundraising and searching for SFI to secure its first permanent home. Before Gell-Mann's passing in 2019, convicted sex offender Jeffrey Epstein, for whom New Mexico was always something of a playground,[44] became chummy with SFI scientists including Gell-Mann,[45] and "sometime prior to 2007"[46] made donations totaling $250,000, as well as a one-off 2010 donation of $25,000, which SFI renounced through re-donation to the Solace Crisis Treatment Center in 2019.[47]

Fortunately for the financial health and repute of the Institute, some of its attendees found practical applications for its lessons in emergent economics, among them, the aforementioned William H. "Bill" Miller III, who serves as the Institute's resident Warren Buffet. Miller "first visited SFI in 1991 and joined its board in 1995. He has since served SFI as Chair, Vice-Chair, and Chair Emeritus of the Board of Trustees," donating $50M to SFI in 2021.[48] Miller credits "external Professor Brian Arthur's work on lock-in technologies and path-dependence" for his early bets on tech investments, then considered risky, and "an early SFI topical meeting on innovation and evolution" to his decision to buy Google during its IPO days.[49] Other meetings inspired him to invest in Amazon and Bitcoin, and the dividends of his stewardship have contributed to the health of SFI's endowments, fellowships, and campuses.

If a financial X-ray reveals the soul and prospects of an institution, SFI currently enjoys a good bill of health, and its new campus ensures another income stream as it offers educational retreats for corporations (think Google execs taking part in a seminar about the latest trends in collective behavior, self-organized criticality, and "consciousness" as manifested in online social networks). Having proven that its concerns were no flash in the pan, SFI can point to a record of institutional successes, including the 2021 Nobel Prize in Physics awarded to Syukuro Manabe, Klauss Hasselmann, and Giorgio Parisi "for groundbreaking contributions to our understanding of complex systems." Parisi's award was "for the discovery of the interplay of disorder and fluctuations in physical systems from atomic to planetary scales" and the committee acknowledged his "mak[ing] it possible to understand and describe many different and apparently entirely random materials and phenomena, not only in physics but also in other, very different areas, such as mathematics, biology, neuroscience and machine learning."[50] As David Krakauer states, "Parisi's work on physical models for complex systems . . . [was] built on the work of SFI researchers including Phil Anderson and David Sherrington."[51] Indeed, both are cited repeatedly in the Nobel Committee's scientific background documentation for the prize,[52] and the list of SFI-affiliated SFI fellows who have received the highest accolades of professional standing grows with each passing year.

At the same time, in its maturity, SFI faces the challenges of its own vindication. What was once considered avant-garde—the frisson of *interdisciplinary science!*—has entered the mainstream, in some quarters, to the point of cliché. Complexity theory, AI, and the many streams of SFI research are staples of science headlines, which can have the ironic effect of diluting SFI's salience, as its original signal competes with growing noise. In such a moment, SFI might like to capture a bit of lightning in a bottle to focus its vision and to shake off the inevitable irrelevance and unacknowledged contributions of middle age. "The design here is to have an intellectually interesting place," Luis Bettencourt explained in 2012. "And the payoff often doesn't come immediately. But it's impossible to do what we do without an environment so rich and diverse and haphazard."[53] True to its center, SFI places long-term bets that the whole will be greater than its parts.

Where SFI has been at the leading edge of relevance we find Cormac McCarthy as well. The two books (the aforementioned duology) that are likely to be his final authorized publications, *The Passenger* and *Stella Maris* read like love letters from a long, mutual intellectual courtship with

SFI. Although elements of *The Passenger* occurred to McCarth,
1962—pause to consider the rarity of a novel sixty years in the making
an indication of the enduring cycles of McCarthy's ruminations—he wo
on completing the duology in fits and starts throughout his tenancy at th
Santa Fe Institute during a period of declining health, and careful readers
will see its influence across the pages of the duology, but especially in *Stella
Maris*.[54] There are layers of mutual reinforcement between SFI's journal,
Nautilus, and Cormac McCarthy's writing. SFI Director David Krakauer
scored a small literary coup by publishing McCarthy's first nonfiction
musings, in no small part the product of their conversation and others at
SFI, on the pages of *Nautilus*.

Perhaps the timing is simply serendipitous, but around a month before
the release of *The Passenger, Aeon* magazine, published an essay "in
association with the Santa Fe Institute, an Aeon Strategic Partner," by
David H. Wolpert, an SFI professor who works in theoretical physics. Titled
"A Sliver of Reality," the think piece asks whether science, mathematics,
and human intelligence will ever be able to capture the smallest part of
reality. McCarthy's new novels treat these questions as a central concern,
and, like Wolpert, they "marvel at the limits of human language, and the
fact that the limitations appear to be universal."[55] Wolpert reflects on the
problem of using finite sequences of symbols to refer to something other
than themselves, the "so-called 'symbol-grounding problem' in cognitive
science and philosophy. The field of mathematics has reacted to this
observation in a similar way, expanding formal logic to include modern
model theory (the study of the relationships between sentences and the
models they describe) and metamathematics (the study of mathematics
using mathematics)." *Stella Maris* is practically obsessed with these issues,
and whether mathematics can ever give rise to a sort of math-beyond-
math, free of the symbol-grounding and indeterminacy problems. Wolpert
ultimately wants to point to "how horribly, and perhaps horrifyingly,
limited and limiting our achievements are—our language, science, and
mathematics."

So do McCarthy's final novels, where one can see the truth of Wolpert's
conclusions and the value of his questions working on him, and the dawning
realization that no matter how well one understands various grammars,
comprehending the tiniest sliver of reality might well be beyond our grasp,
as the grammar itself reveals the bounded limits of our perception. "Taken
together," Jonathan Miles concludes in an enthusiastic advance review of
the duology in *Gun and Garden*,

> *The Passenger* and *Stella Maris* are an intellectually breathtaking
> achievement, an electric and thunderous attempt, as the Thalidomide
> Kid says to Bobby to "get hold of the world." Not that such is possible,
> according to the Kid. No, "you can only draw a picture;" he says. "Whether

it's a bull on the wall of a cave or a partial differential equation—or an astonishing pair of novels—'it's all the same thing.'"[56]

In other words, even when we shut up and do the math, per the famous dictum, we find ourselves bounded by a small set of *known* cognitive constructs, and the more that science reveals of the so-called known universe, the better we understand our trifling place in the firmament of the infinite. Along with our smallness comes the latent realization of the smallness of what we can see.

On some level, I felt this is my bones when I crossed the threshold at SFI, an institution that has little reason to talk to me, a career humanist for the most part. Cormac writes in the Institute's unofficial manifesto, "Occasionally we find that an invited guest is insane. This generally cheers us all up. We know we're on the right track." Per Cormac's litmus test, I cannot claim to "know more than anybody else about a subject,"[57] so I suppose that left me with the recourse of fox-craziness to slip through the gates. Ultimately, it was the kindness of the McCarthy family, the generosity of the Institute in suffering pests, and persistence that yielded a chance to peer inside the place. After decades in higher education settings, I've often been buoyed by the occasional madness and brilliance of colleagues, yet I can honestly say that even when SFI was largely ghosted by the pandemic, it crackled with a unique Great Group energy. I spent the day dancing on the toes of two right-brained feet, trying to keep up as we tangoed across disciplines, books, conversations, and ideas. David Krakauer and Tim Taylor were superbly versed in literary history and the history of ideas, broadly; by comparison, my knowledge of the history of science was scanty and patchwork.

I had breached reciprocity in another way. When I arrived, Taylor rattled off from memory an in-depth version of my professional bona fides, astonishingly well researched and accurate ("all public record" he was quick to add). I burned with the shame of the A-student who has not done his homework, for what did I know of SFI? My understanding of complexity science had largely foundered on a few attempts to read relevant mathematical and scientific treatises. (That day it was Taylor who gently steered me to the Institute's own Geoffrey West's magisterial *Scale: The Universal Laws of Growth, Innovation, Sustainability, and the Pace of Life in Organisms, Cities, Economies, and Companies* [2018] as a point of entry to complexity theory, which proved to be a revelation.[58]) I was disappointed with myself for wandering in cavalierly as an ingenue and ignoring the truism that chance favors the prepared mind.

The only other disappointment of that day resulted from a miscommunication. Early on, I asked if I might take some notes by way of photos and brief recordings with my phone. Taylor misunderstood my question to be journalistic in nature, perhaps a prelude to a formal interview. He preferred to stay off record, and I did not press the point, leaving a once-in-a-lifetime tour to my memory, which is almost as porous as my understanding of complexity theory. I have patched some of the gaps with bits of interviews and contextual material that capture the spirit of the day's discussions, if not the letter.

Afterward I felt called to a more profound sort of intellectual honesty and discipline as a way of paying down the gift of that day. At best, I could make some sense of what I had seen, heard, and sensed, to give other nonspecialists a sense of that electricity. Dennis McCarthy once told me that Santa Fe held no special charm for Cormac outside of the Institute. Cormac McCarthy had elected to make it his home, even when, for the well versed, the intellectual work of understanding the place could be overawing. As his friend, writer/actor Sam Shepard, said with his usual louche delivery, apart from him and Cormac, "Everybody else [at SFI] is a nuclear physicist. Which is cool, you know. But it leads to a lot of conversational dead-ends."[59]

Not so for McCarthy, who, for all his shyness, is often credited as a conversational emcee at SFI. As Allison Flood wrote in *The Guardian*, McCarthy's "knowledge of physics and maths" purportedly exceeds "that of many professionals in the field," and he is not just a conversant but fluent and valued partner in SFI discussions.[60] "He regularly attends the workshops at S.F.I.," Richard Woodward wrote in *Vanity Fair*,

> where the topic may be the evolution of prion proteins or mammalian muscle adaptations or lying and deception or bounded inferences for decision-making in games. As a result, he often serves as a clearinghouse for those who want to know what everyone else is up to. "I find it easier to talk to Cormac about what J. Doyne Farmer is doing"—Farmer is the economist-physicist-gambler celebrated in Thomas A. Bass's best-selling book *The Eudaemonic Pie*—"than to talk to Doyne himself," says the Russian linguist Sergei Starostin.[61]

Additionally, "[a]mong this rarefied gathering of leading intellects, none is more respected than the spry old cowboy dipping his tortillas in beans at the lunch table," David Kushner writes. "Dressed in a crisp blue shirt and jeans, he sits comfortably with his weathered boots crossed and listens intently as a theoretical biologist who has flown in from Berlin discusses something called evolutionary economics—the relationship between animal behavior and marketlike forces. This is quintessential Santa Fe stuff, examining one phenomenon (biology) in the light and lexicon of another (economics)."[62] McCarthy learned Spanish, the better to write the Border Trilogy, and at SFI he read scientific papers in their original language, so to speak, patiently puzzling out mathematical problems so that he might understand their details as well as the underlying history of mathematical theory. Alicia says in *Stella Maris*, "Verbal intelligence will only take you so far. There is a wall there, and if you dont understand numbers you wont even see the wall. People from the other side will seem odd to you" (69). In this, I hear Cormac's determination to climb the wall, if only for the sake of scaling another one. SFI's residents see Cormac as the man in the middle because he does his homework, fastidiously absorbing the work of his colleagues, literally doing the math to understand it (and, as we know from studies, math is infamously mentally effortful such that for some it is neurophysically indistinguishable from other experiences of pain).[63]

When I visited, I soon had a sense of the place's interdisciplinary fluency and collegiality, and how McCarthy would find it so convivial. Adepts in translation, David Krakauer and Tim Taylor accommodated a wide-ranging discussion of contemporary and classical literature, favorite big idea science books, politics, and theory, even as they steadily downloaded the history of the Institute and its work. The pace of it was breathless and energetic, free-spirited and wide-ranging, *My Dinner with Andre* with a philosophy of science slant. My small conversational contribution was to describe my recent research in disinformation applying epidemiological theory, drawing parallels to emergent systems that produce black swan events. "If you ever need someone for a Division of Artificial Stupidity, sign me up," I joked.

Of course, this idea is hardly new to SFI. In a recent interview, Krakauer said,

> I've always been interested in living phenomena. What makes life different from non-life, what makes the biological world distinct from the physical world? And one question is What is life? and another question is What is intelligent life? And is that distinction real? So I guess that's been my career trying to understand what life is, what biology is, what culture is and how it manifests either intelligently or foolishly or stupidly.[64]

In a 1993 interview with the *Santa Fe New Mexican*, Gell-Mann, "beaming ... talk[ed] about 'frozen accidents,' chance events that occur in the evolution of a complex adaptive system–like world history–that have far-reaching ramifications."[65] The subtext is that Gell-Mann and other members of the early SFI used simplicity-complexity complementarity to change the course of world history with the promise and terror of the nuclear age. "Perhaps in the world's unmaking it would be possible to see how it was made," McCarthy writes in *The Road*, which is perhaps the most explicit reference to be found in his opus regarding SFI's interest in evolutionary complex systems.[66] While SFI's mission has not been expressly geared to peace studies or putting the nuclear djinni back in the bottle, it has been very interested in disaster complexity events and their containment. "Over the years," we learn on the SFI website, "Laurence Gonzales never lost his interest in science, nor in questions stemming from airline crashes, such as: Why do smart people do stupid things?"[67] There is an unspoken allegorical subtext: Can those laboratory wonks who invented the potentially world-ending compound, Ice Nine, in Kurt Vonnegut's not-very-fictional *Cat's Cradle* (1963) find theoretical ways to mitigate the problems they have created, both seen and unforeseen?

McCarthy himself recognizes the difficulty of scientific translation. In fact, he has dedicated the last two decades of his writing career (or perhaps phased retirement) to it, specifically addressing in his own work the topic of what constitutes strong scientific writing. He line-edited Gell-Mann's *The Quark and the Jaguar* (1994), helped Lisa Randall complete *Knocking on Heaven's Door: How Physics and Scientific Thinking Illuminate the Universe and the Modern World* (2011), copy-edited Lawrence Krauss's biography of Richard Feynman, *Quantum Man* (2012), and, in 2017, surprised many with his first Kekulé essay, a foray into publishing his own scientific nonfiction. The 2022 duology "references renowned mathematicians and physicians such as Werner Heisenberg, Robert Oppenheimer, Wolfgang Ernst Pauli, Albert Einstein, Richard Feynman, James Clerk Maxwell, Pierre-Simon Laplace, Johann Carl Friedrich Gauss, Leonhard Euler, and John von Neumann."[68]

For models of the conversation between science and humanities we can begin by observing McCarthy's passion for science's intellectual ferment. Again, what Bennis says of great groups applies to Cormac's milieu at SFI:

> Curiosity fuels every Great Group. The members don't simply solve problems. They are engaged in a process of discovery that is its own reward. Many of the individuals in these groups have dazzling individual skills—mathematical genius is often one. But they also have another quality that allows them both to identify significant problems and to find creative, boundary-busting solutions rather than simplistic ones. They have hungry, urgent minds. They want to get to the bottom of everything they see. Many have expansive interests and encyclopedic knowledge.[69]

The exhaustion of language seems only to have drawn McCarthy to the alternative language of mathematics. As Alicia says in *Stella Maris*, "Intelligence is numbers. It's not words. Words are things we've made up. Mathematics is not. The math and logic questions on the IQ tests are a joke." When her psychiatrist asks her, "How did it get that way?" she replies, "Maybe it always was. Or maybe we actually got there by counting. For a million years before the first word was ever said (19)." McCarthy's childhood fascination with trapping, fossils, and taxonomy can be detected in the fact that "his knowledge of the natural world is vast and includes many of the Latin names of birds and animals. He can discourse on Harris's hawks ('the only raptor that hunts communally') or on poker (Betty Carey, the former high-stakes player, is an old friend) or on how gun manufacturers rifled their barrels before the invention of metal lathes."[70]

The point is not that McCarthy is a polymath, which, after all, could reflect a type of common insecurity, or that he is more to be admired as a writer because he knows mathematical history and taxonomy. Thomas Halliday recently argued in the SFI-affiliated journal *Nautilus* that taxonomy is overrated in any case: "Ever since the earliest attempts at classifying the natural world, humans have been labeled separately from the rest of life, as something apart, something special. The trouble with taxonomic labels is that they, like communities of organisms, are not constant through time"— or, in another word, emergent.[71]

The point is not that Cormac McCarthy is special, or even a special human, but, rather, the beauty to be discovered in chasing all of these modes of curiosity, even though, as McCarthy said during an interview at SFI, "It's hard to define beauty."[72] These connections bring us finally to the possibility of understanding that indeed there exists what Cormac McCarthy has called "an aesthetic of science." In proposing the existence of such an aesthetic, McCarthy commented on the profound links between the excitement of scientific discovery and powerful writing: "Both involve curiosity, taking risks, thinking in an adventurous manner, and being willing to say something 9/10ths of people will say is wrong."[73]

How are we to make sense of the proposition that, having labored to achieve depth of field within a native discipline, it must be surrendered to the deeply defamiliarizing position of another? Perhaps because curiosity itself is the object, and variety is the life of spice. Has literary theory become a sort of terminal Core Theory in literary studies, and does it provide the intellectual tools to refresh curiosity and discovery among nonspecialists? Or have these tools—in concert with algorithm-driven electronic research that makes only certain things salient and that leads to convergent, unoriginal conclusions—impinged on discovery altogether, evolved a set of shopworn interpretive conventions, and substituted data for theory by insinuating an artificial intelligence within our own? David Krakauer has his doubts.

> I often think about word processors, for example. Everyone is using the same one and it is dictating the forms that channel your ideas. You could say, *it doesn't matter, I mean, everyone's using the same typewriter the turn of the century the last century—that didn't limit the creative potential of a Philip Roth or a Saul Bellow or a Cormac McCarthy.* But something feels a little different about this time, especially with AI, because it starts to tell you what you should do and think, this is how you should present your ideas well, this is the correct grammatical construct, oh and by the way here are some ideas from Wikipedia that bear on the narrative that you're creating. So there are forces of homogeneity in play that might have not been present in earlier technologies.[74]

Perhaps this is true of latest science as well. Recently, SFI Fractal Faculty member and External Professor Sean M. Carroll from the California Institute of Technology suggested that current models in physics are *good enough* for our quotidian purposes. The materialist conclusions that flow from the best equation we have "on which the everyday world supervenes" (more on this later) might read like troubling articles of faith:

Assuming Carroll's claim is correct, it has a number of immediate implications. It means there is no life after death, as the information in a person's mind is encoded in the physical configuration of atoms in their body, and there is no physical mechanism for that information to be carried away after death. The problems of consciousness must ultimately be answered in terms of processes that are compatible with this underlying theory. And while historically, discoveries of new particles and forces have spurred technological innovations, Core Theory means that won't happen going forward, since those discoveries won't be at a level to impact our everyday lives.[75]

Perhaps this might be set against the spiritual yearning of even a scientific realist. In Cormac's words,

I have a great sympathy for the spiritual view of life, and I think that it's meaningful. But am I a spiritual person? I would like to be. Not that I am thinking about some afterlife that I want to go to, but just in terms of being a better person. I have friends at the Institute. They're just really bright guys who do really difficult work solving difficult problems, who say, "It's really more important to be good than it is to be smart." And I agree it is more important to be good than it is to be smart. That is all I can offer you.[76]

In arguing that science and art are connected by aesthetics, this study can only confirm the profound truth of McCarthy's unwavering belief that "There's a beauty to science," and a language of human understanding that transcends words and that might even be held in contempt of them.

* * *

Each chapter in this study has a different genesis, some many years in the making. For example, the journey into the astonishing regressions and expansions of chirality (Chapter 2) began with a detail from the opening passage of *Suttree*, prompting me to ask, with childlike persistence, a series of *why* questions to any expert patient enough to offer a reply: Why do so many living things favor asymmetry in a certain direction? Is it that way everywhere? What causes the bias? Why do compounds tip that way? And so on. Following the whys through the hall of mirrors led me to the limits of what is known about the origin of things and to long-forgotten lessons in organic chemistry. The chapter owes much of its substance to colleagues who enlarged my understanding of the evolutionary biology of chirality. But its origin is ultimately in McCarthy's scientifically observant eye for detail—indeed, a lowly plant mentioned *passim* on one of his pages—and his ability to recast the prosaic world in the vivid light of literary imagination.

McCarthy's few full-length published interviews and his appearance on the Oprah Winfrey Show occurred during his SFI years and offer resources for understanding what SFI means to him—if taken at his word or that of his interpreters. Nick Romeo's much-remarked 2012 *Newsweek* interview found McCarthy close to giddy about his new home away from home in prime days at the Institute—still a fresh chapter of life in a "new" town, with a new batch of colleagues, a newish son and family life. Of course, the story is considerably more complex, as usual, than condensate of *Newsweek*. What, I wondered, did the place mean to him and his work with the passing of another decade, and how was his life intertwined with the Institute's story? As I started my research, however, I learned that Laurence Gonzales was writing an extensive history of SFI, one proposed to be more comprehensive and up-to-date than Mitchell Waldrop's SFI-focused, sturdy-but-faded book *Complexity: The Emerging Science at the Edge of Order and Chaos* (1993), written before SFI marked its first decade.[77]

A visiting Miller fellow at SFI, Gonzales's life is fascinating in its own right. He has worked alongside McCarthy and in the Institute for many years, often adding to his history of the Institute in real time as a first-person observer. A musician who wandered for a time with Ken Kesey, he toiled as an award-winning freelance writer for much of his career, probing topics such as the complexity of disaster situations and their reconstruction.[78] In summer 2021 Gonzales graciously offered to speak with me about his work. We discussed both the necessity of an interior view to do justice to SFI's enterprise and the difficulty of sifting through patchwork accounts of the Institute resulting from the lack of an authoritative history—a gap that his work might yet fill. The Institute itself has been more forthcoming about its own history lately, offering the first installment of something like an authorized history on its webpages. This information once was difficult to find. With Gonzales intending to offer a fine-grained account of Cormac's involvement through the years in SFI, and with new information coming online as SFI ripens into its middle age, including chronicles on its own webpages, there is less need to replicate or recapitulate SFI's history here.

Instead, my brief will be short and narrow with a just-the-facts overview of Cormac's involvement, a chronology that begins with the official history from SFI: "Since the nineties McCarthy has been a fixture at the Santa Fe Institute."[79] (Cormac-the-Word-Authority, a chip from his father's block, might find some amusement there, given that fixtures in real estate convey with property.) Exactly when in the 1990s he became a fixture is hard to ascertain without access to primary records; he was likely an occasional visitor throughout the decade, becoming a Santa Fe resident around 2000. He married Jennifer Winkler in "early 1998," attended SFI's "Complexity and Simplicity" Symposium in December of that year,[80] and in 1999, the film adaptation of *All the Pretty Horses* was shot largely around Santa Fe.[81]

He was probably not fully integrated into SFI residency there, formally or informally, until sometime around 2001.[82] In any case, McCarthy is listed as an SFI trustee for the first time in SFI's 2011 tax filing, about a decade after settling into the community there.[83]

To set his move to Santa Fe in spring of 2001 within the larger scope of McCarthy's life, he started going by Cormac forty years earlier; published his first book thirty-five years earlier; first decamped for the southwest around twenty years earlier; and first met Gell-Mann around twenty years earlier. This retrospect underscores the point that McCarthy's SFI phase would be, for more conventional souls, called retirement. For McCarthy, the ideal unretirement home happened to be an academic Shangri La, a palace of ideas, a place where

[l]unchtime conversations range from game theory to historical linguistics to Sophocles. Pulitzer Prize-winning authors, Nobel Laureates and MacArthur geniuses wander the halls, scrawling equations on the windowpanes with erasable markers. The novelist and philosopher Rebecca Goldstein calls it "everything I hoped academia would be as a graduate student." She adds, "It was pure bliss."[84]

If a primary advantage of Santa Fe for Cormac was to make the Institute his study, social club, and office, a secondary advantage was that it happened to be a place where Cormac's brother, Dennis, and his wife, Judy, had long dreamed of retiring. And as McCarthy made the rounds during working days at SFI, he would have received an education in complexity science and emergence by osmosis only reinforced by his preexisting interest. Ciarán Dowd submits that "McCarthy's long immersion in the conceptual vocabulary of his friends and acquaintances at SFI will doubtless have inculcated within him a familiarity with the concept of emergence," which, "as a philosophical concept . . . has existed for over a century," but, as has been discussed, has had a relatively short existence as an acknowledged field of scientific inquiry, with one of its most significant nodes at SFI.[85] In sussing out McCarthy's relationship to the Institute, Dowd parses the various roles that McCarthy has played: all-around booster/influencer, occasional fundraiser, a reliable connector for conversations between kindred spirits, an SFI-inspired published respondent through his Kekulé essays, and a frequent participant in the discourse surrounding "contingency and chance" as well as linguistics. As to the Institute's influence on McCarthy, Dowd asks, "can we find these 'covert manifestations,' as [SFI director David] Krakauer called them, of the author's scientific interests?" Surveying McCarthy's publications from the SFI years, Dowd returns with a resounding "yes." He suggests, for example, that *The Road* "hints at one possible resolution of McCarthy's careerlong metaphysical tension between naturalism and supernaturalism, a resolution suggested by the kind of 'complexity thinking' current among those working at the Santa Fe Institute."

This might be matched to Dowd's first area of influence, which, he claims, lies in McCarthy's prose itself: "within a paratactic syntactical frame, McCarthy's prose runs the spectrum between two registers: one flat and declarative, and of a piece with the syntax; the other heightened and exceedingly figurative, straining against the confines of the paratactic frame."[86] In other words, Dowd points to a trend in McCarthy's prose, which Allen Josephs and others have mapped in terms of a tension between his earlier Faulknerian and later Hemingwayesque styles, that itself might nod to the evolution of a prose style informed and enhanced by scientific understanding. Dowd's chapter epigraph is from McCarthy's interview in the documentary *The Unbelievers* (2013): "People who like science are all drawn to it for the same reason—it explains the physical world: 'What is this stuff?' Is it the last word? On what reality is, on what the physical world is? I don't know, but if it's not the last word it's at least the best word."[87]

I would frame it this way: having come to the edge of the literary limits of a lifelong preoccupation with questions related to complex systems, McCarthy set himself to grokking its expression in mathematics and science. As early as 1978 he was writing to Guy Davenport about Heisenberg's theory of indeterminacy.[88] Almost forty-five years later we have his conclusion, laid out in the final diptych comprising the novels *The Passenger* and *Stella Maris*, that complexity science may have the first word, but neither of the dialectics he had sought to master (science or creative narrative) has the last. Complexity science, for McCarthy, is a means to speak of emergent universes with more precision. Just as contemporary philosophy is largely framed in the symbols of logic and mathematics—some would say, *is* mathematics— in its quest to find better descriptive systems, McCarthy's search for the best-known responses to his metaphysical questions that would distinguish between what is, physically, and what simply exists without a physical reality, necessarily required his residency and time in the eaves of the Santa Fe Institute.

I would submit also that McCarthy's earnest pursuit of more and better answers to the problems that were ineffable to him from childhood offers a fine model of a dream of life conducted along its course by curiosity itself. This is the work of intellectual reconciliation, accepting that our small knowing is rooted at once in infinite recursion and expansion, a quest for emergent logics of approximation. Following a similar path, Timothy Andersen, principal research scientist at Georgia Tech Research Institute, found himself stymied by the problems of mathematical inconsistency posited by the likes of Alfred North Whitehead and Bertrand Russell, prompting him to switch from mathematics to physics, only to hit another wall (Alicia reviews some of the same historical figures and the way to Gödel's incompletion theorem via Wittgenstein in a handful of pages of dialogue in *Stella Maris*).[89] "Quantum theory exposed that, too, as a fantasy," he wrote, "even though we could define rules and equations for physical laws, we could not explain what they meant. Recent experiments in quantum information theory have shown that our most basic assumptions about reality, such as when something can be considered to have been observed and to have definite physical properties, are in the eye of the beholder."[90]

Andersen lays out the intellectual dimensions of the quandary in plain terms, and without going into the details (read his article, which takes on the double-slit experiment and the surprising problems physicists face in observing waves), suffice it to say that the German philosopher Ludwig Wittgenstein wrestled with the same problems of indeterminacy. As Andersen puts it, Wittgenstein showed that the "realist versus anti-realist debate is meaningless because both sides are trying to *say* things that are only *showable*," which has a bearing, for example, on the limits of Effective Field Theory, the current paradigm "underlying modern theoretical physics" that fashions a hybrid Core Theory from "the Standard Model of particle physics plus Einstein's general relativity."[91] In short, scientific inquiry proceeds from a sort of tautology circumscribed by the limits of scientific realism. We might wish that mathematics should describe an underlying reality instead of supervening on it. As Krakauer has stated,

> We still are living with Einstein's general theory and the theory of quantum mechanics that have not been reconciled for a hundred years and we have all those tools, we can deploy them effortlessly at almost zero cost, but no one has had an idea that really has taken us much further than they did a hundred years ago, so I think the real discussion has to do with the quality of what you produce.[92]

In a recent paper, SFI Fractal Faculty member and External Professor Sean M. Carroll from the California Institute of Technology acknowledges the stalemate as good enough for government and most other work bounded by earth's energy scales: "Effective Field Theory (EFT) is the successful paradigm underlying modern theoretical physics, including the 'Core Theory' of the Standard Model of particle physics plus Einstein's general relativity." He argues that

> EFT grants us a unique insight: each EFT model comes with a built-in specification of its domain of applicability. Hence, once a model is tested within some domain (of energies and interaction strengths), we can be confident that it will continue to be accurate within that domain. Currently, the Core Theory has been tested in regimes that include all of the energy scales relevant to the physics of everyday life (biology, chemistry, technology, etc.). Therefore, we have reason to be confident that the laws of physics underlying the phenomena of everyday life are completely known.

If anyone asks you to show the math, keep this formula in your pocket:

$$A = \int_{k<\Lambda} [Dg][DA][D\psi][D\Phi] \exp\left\{ i \int d^4x \sqrt{-g} \left[\frac{1}{16\pi G}R - \frac{1}{4}F_{\mu\nu}F^{\mu\nu} + i\bar{\psi}\gamma^\mu D_\mu\psi \right.\right.$$

$$\left.\left. + |D_\mu\Phi|^2 - V(\Phi) + (\bar{\psi}_L^i Y_{ij}\Phi\psi_R^j + \text{h.c.}) + \sum_a \mathcal{O}^{(a)}(\Lambda) \right] \right\}$$

FIGURE 1.3 *Core Theory, Equation 7, from SFI Affiliated Professor Sean Carroll's* "The Quantum Field Theory on Which the Everyday World Supervenes."

For some, the equation is a thing of beauty, and for others, about as satisfying as the number 42, worked out as the "Answer to the Ultimate Question of Life, the Universe, and Everything" by the Deep Thought supercomputer over the course of 7.5 million years in Douglass Adams's *Hitchhiker's Guide to the Universe*. A parallel can be drawn to Cormac's dialectics of narrative and math/science, and the idea that the last word is that there not only is no last word, but there *can also be* no last word, no theory of everything to be devised. "As a scientist and mathematician," Andersen writes, "Wittgenstein has challenged my own tendency to seek out interpretations of phenomena that have no scientific value—and to see such explanations as nothing more than narratives." It is not novel to say that science and narrative are conjoined and vary by means of expression, but it is interesting to consider the principled life-journey, a literalistic and hard-nosed search to make sense of the biggest questions (what is here, and why are we here) that ultimately brought McCarthy to a place like Andersen. "I have humbled myself to the fact that we can't justify clinging to one interpretation of reality over another," Andersen writes. "In place of my early enthusiastic Platonism"—something he seems to share with McCarthy—he has come "to think of the world not as one filled with sharply defined truths, but rather as a place containing myriad possibilities–each of which, like the possibilities within the wavefunction itself, can be simultaneously true."[93]

In sum, SFI offers a new way of conceiving of McCarthy's creative career in the process. Currently scholars use the bookends of his Appalachian/ Faulknerian years (from *The Orchard Keeper* to *Suttree*), the ensuing southwestern phase (*Blood Meridian* and the Border Trilogy, reflecting a middle place in the transition from a Faulknerian to a Hemingwavian style), and what I would term, finally, his Science/SFI phase, spanning the questions of determinism announced in *No Country for Old Men*, and apotheosized in the final duology of *The Passenger/Stella Maris*. David Krakauer himself suggested that "after McCarthy went through an 'Appalachian phase' and a 'Southwestern phase,' the new book [duology, as it turns out] is going to be 'full-blown Cormac 3.0, a mathematical-analytical novel.'"[94] It is not so much a new departure as reaching a long-sought waypoint, not for disembarkation, but for a life spent in continual wonder of expanding universes.

Notes

1   Richard B. Woodward, "Cormac Country," *Vanity Fair*, August 1, 2005, https://www.vanityfair.com/culture/2005/08/cormac-mccarthy-interview.

2   Leonard Bernstein, *The Unanswered Question: Six Talks at Harvard* (Cambridge, MA: Harvard University Press, 1976), 3.

3   "Bill Miller Funds SFI Expansion," The Santa Fe Institute, January 11, 2018, https://www.santafe.edu/news-center/news/bill-miller-funds-sfi-expansion.

4   Cf., James Somers, "The Man Who Would Teach Machines to Think," *The Atlantic*, November, 2013, https://www.theatlantic.com/magazine/archive/2013/11/the-man-who-would-teach-machines-to-think/309529/.

5   McCarthy talks about his friend George Zweig as well as Gell-Mann in the trailer for *Desert Shift,* directed by Karol Jalochowski, n.d., https://www.youtube.com/watch?v=Dw8Ku3m1rXo, and at length in "couldn't care less," https://www.youtube.com/watch?v=HrUy1Vn2KdI.

6   "Complexity and History," The Santa Fe Institute, https://www.santafe.edu/research/themes/complexity-and-history; John H. Miller, *A Crude Look at the Whole: The Science of Complex Systems in Business, Life, and Society* (New York: Basic Books, 2015).

7   "What is SFI?" The Santa Fe Institute, https://www.santafe.edu/about/faq.

8   Nick Romeo, "Cormac McCarthy on the Santa Fe Institute's Brainy Halls," *Newsweek*, February 12, 2012, https://www.newsweek.com/cormac-mccarthy-santa-fe-institutes-brainy-halls-65701.

9   Brian Williams, "Glyphic Alphabet," https://parasol.io/Glyphic-Alphabet.

10 "Study: Complexity Holds Steady as Writing Systems Evolve," The Santa Fe Institute, June 16, 2021, https://www.santafe.edu/news-center/news/study-complexity-holds-steady-writing-systems-evolve.

11 "Bill Miller funds SFI Expansion."

12 Bob Quick, "S.F. Institute Gets Zoning Exception," *Santa Fe New Mexican*, April 24, 1993, 13.

13 Keith Easthouse, "New Home for Institute," *Santa Fe New Mexican,* August 21, 1994, E2.

14 James G. McGann, "2020 Global Go to Think Tank Index Report," University of Pennsylvania, January 28, 2021, https://repository.upenn.edu/think_tanks/18/, 198, 280.

15 "Miller Campus," The Santa Fe Institute, https://www.santafe.edu/culture/miller-campus.

16 Absent very auspicious circumstances (e.g., sheltered calderas), promontories are rarely conducive to habitation. Game and humans largely avoid them because by nature they are effortful to reach, buffeted by the more severe reaches of the elements, places where water more often begins than collects.

17 "Gift of Property to SFI Adds a Quiet Setting for Scholarly Thought," Santa Fe Institute, February 11, 2013, https://www.santafe.edu/news-center/news/tesuque-campus-announce.

18 Stephen Wolfram, "My Part in an Origin Story: The Launching of the Santa Fe Institute," *Stephen Wolfram Writings* (blog), June 18, 2019, https://writings .stephenwolfram.com/2019/06/my-part-in-an-origin-story-the-launching-of-the -santa-fe-institute/.

19 Ibid.

20 Ibid.

21 Ibid.

22 "Engage with SFI," The Santa Fe Institute, https://www.santafe.edu/engage.

23 Cormac McCarthy, "SFI's Operating Principles," The Santa Fe Institute, 2017, https://www.santafe.edu/about/operating-principles.

24 Romeo, "Cormac McCarthy on the Santa Fe Institute's Brainy Halls."

25 McCarthy, "SFI's Operating Principles."

26 See, for example, Derek Thiess's *ISLE* article, "On The Road to Santa Fe: Complexity in Cormac McCarthy and Climate Change"; passing discussions of science in Stephen Frye's *Understanding Cormac McCarthy*; and the chapter devoted to McCarthy in Gordon Slethaug's *Beautiful Chaos: Chaos Theory and Metachaotics in Recent American Fiction*.

27 Cf., Sean Braune, "A Chaotic and Dark Vitalism: A Case Study of Cormac McCarthy's Psychopaths Amid a Geology of Immorals," *Western American Literature* 50, no. 1 (2015): 1–24. I am indebted to Jackson Kulick not only for pointing to some of these examples but also for sharpening my understanding of the main strands of debate surrounding quantum mechanics.

28 Jean Bricmont, "Determinism, Chaos, and Quantum Mechanics," (n.d.), http:// www.freeinfosociety.com/pdfs/mathematics/determinism.pdf; Jean Bricmont, *Making Sense of Quantum Mechanics* (New York: Springer, 2016).

29 See Persi Diaconis, Susan Holmes, and Richard Montgomery, "Dynamical Bias in the Coin Toss," *SIAM Review* 49, no. 2 (2007): 211–35.

30 Ciarán Dowd, "The Santa Fe Institute," in *Cormac McCarthy in Context*, ed. Steven Frye (Cambridge: Cambridge University Press, 2020), 33–44, doi: https://doi.org/10.1017/9781108772297.005; Lydia R. Cooper, *Cormac McCarthy: A Complexity Theory of Literature* (Manchester: Manchester University Press, 2021).

31 Cooper, *Cormac McCarthy*, 12, 18, 7, 8.

32 Cormac McCarthy, "Connecting Science and Art," *Science Friday* interview by Ira Flato, National Public Radio, April 8, 2011, https://www.npr.org/2011/04 /08/135241869/connecting-science-and-art.

33 "About Us," *Nautilus*, https://nautil.us/about-us/.

34 For example, the chapter in this volume on chirality, the quality of the handedness of things ranging from molecules to plants to stars, was over five years in the making, requiring consultation with scientists from multiple fields, philosophers and theologians, and, yes, creative writers attuned to science.

35 Fortunately, I have such a friend, and I am grateful to Jackson Kulick for shedding light on this, and to Adrian Rice, whose infectious enthusiasm fired my interest in the history of math generally.

36 Cormac McCarthy, "Cormac McCarthy Returns to the Kekulé Problem," *Nautilus,* November 27, 2017, https://nautil.us/cormac-mccarthy-returns-to -the-kekul-problem-6832/.

37 Letter from Cormac McCarthy to Guy Davenport postmarked November 20, 1984, Guy Davenport Papers, Letters from Cormac McCarthy, 1968–1989 and undated, Container 133.2, Manuscript Collection MS-4979, Harry Ransom Center, The University of Texas at Austin.

38 Elizabeth A. Harris, "Early Cormac McCarthy Interviews Rediscovered," *New York Times*, September 30, 2022, https://www.nytimes.com/2022/09/30/books/ early-cormac-mccarthy-interviews-rediscovered.html.

39 Warren G. Bennis and Patricia Ward Biederman, *Organizing Genius: The Secrets of Creative Collaboration* (New York: Basic Books, 2007), xv, 4, 15.

40 John German, "Conception to Birth," The Santa Fe Institute, n.d., https://www .santafe.edu/about/history.

41 Keith Easthouse, "Evolution, Revolution, Accidents, and Murray Gell-Mann," *Santa Fe New Mexican*, August 21, 1994, E-2.

42 "Santa Fe Institute," ProPublica Nonprofit Explorer, https://projects .propublica.org/nonprofits/display_audit/11115520201.

43 German, "Conception to Birth."

44 Matt Farwell, "Jeffrey Epstein Chose New Mexico for a Reason," *The New Republic*, August 15, 2019, https://newrepublic.com/article/154761/jeffrey -epstein-zorro-ranch-new-mexico-history.

45 Neel V. Patel, "Jeffrey Epstein Liked Palling Around with Scientists—What Do They Think Now?" *The Verge*, July 13, 2019, https://www.theverge.com/2019 /7/13/20692415/jeffrey-epstein-scientists-sexual-harassment.

46 Steve Terrell, "Santa Fe Institute Plans to Donate Epstein Money," *Santa Fe New Mexican,* September 17, 2019, A6.

47 "SFI Gives $25K to Solace Crisis Treatment Center," The Santa Fe Institute, December 2, 2019, https://www.santafe.edu/news-center/news/sfi-gives-25k -solace-crisis-treatment-center.

48 "Santa Fe Institute Receives $50 Million from Bill Miller," Santa Fe Institute, November 9, 2021, https://www.santafe.edu/news-center/news/santa-fe -institute-receives-50-million-bill-miller.

49 "Bill Miller Inaugurates Namesake Campus," Santa Fe Institute, November 25, 2019, https://www.santafe.edu/news-center/news/bill-miller-inaugurates -namesake-campus.

50 "Press Release: The Nobel Prize in Physics 2021," October 5, 2021, https:// www.nobelprize.org/prizes/physics/2021/press-release/.

51 "Santa Fe Institute Receives $50 Million from Bill Miller."

52 "Scientific Background on the Nobel Prize in Physics 2021," The Nobel Committee for Physics/The Royal Swedish Academy of Sciences, October 5, 2021, https://www.nobelprize.org/uploads/2021/10/sciback_fy_en_21.pdf.

53 Romeo, "Cormac McCarthy on the Santa Fe Institute's Brainy Halls."

54   Dianne C. Luce, *Embracing Vocation: Cormac McCarthy's Writing Life, 1959–1974* (Columbia, SC: University of South Carolina Press, 2022), 14: "What is clear, then, is that McCarthy partially conceived his long-awaited work *The Passenger* before May 1962 and that he was mulling the project intermittently while finishing *The Orchard Keeper* and drafting *Outer Dark, Child of God,* and *Suttree . . .*"

55   David H. Wolpert, "A Sliver of Reality," *Aeon*, September 5, 2022, https://aeon.co/essays/ten-questions-about-the-hard-limits-of-human-intelligence.

56   Jonathan Miles, "Double Punch," *Garden & Gun*, October/November 2022, 46.

57   McCarthy, "SFI's Operating Principles."

58   Geoffrey West, *Scale: The Universal Laws of Life, Growth, and Death in Organisms, Cities, and Companies* (New York: Penguin Publishing Group, 2018).

59   Robert Nott, "'He Kept Digging, Kept Searching,'" *Santa Fe New Mexican*, July 31, 2017, https://www.santafenewmexican.com/news/local_news/he-kept-digging-kept-searching/article_ac7dfb3d-594e-5630-89d7-f4426f81f922.html.

60   Allison Flood, "Cormac McCarthy's Parallel Career Revealed—As a Scientific Copy Editor!" *The Guardian,* February 21, 2012, https://www.theguardian.com/books/2012/feb/21/cormac-mccarthy-scientific-copy-editor.

61   "Sam Shepard on American Culture, the Stage and Screen, and Writing at SFI," Santa Fe Institute, September 17, 2014, https://www.santafe.edu/news-center/news/guardian-sam-shepard-writing-culture.

62   David Kushner, "Cormac McCarthy's Apocalypse," *Rolling Stone,* December 27, 2007, 43–53, http://www.davidkushner.com/article/cormac-mccarthys-apocalypse/.

63   Jeremy Berlin, "Math Can Be Truly Painful, Brain Study Shows," *National Geographic,* November 7, 2012, https://www.nationalgeographic.com/science/article/121108-math-pain-hurts-brain-science-health.

64   *Digital Transformation: Interview with David Krakauer*, directed by Manuel Stagars (August 2017), streaming documentary video, https://digitaltransformation-film.com/david-krakauer-santa-fe-institute/.

65   Easthouse, "Evolution, Revolution, Accidents, and Murray Gell-Mann."

66   "Laurence Gonzales Named an SFI Miller Scholar," Santa Fe Institute, March 21, 2016, https://www.santafe.edu/news-center/news/gonzales-miller-scholar-announce.

67   Ibid.

68   "One Reporter's Account of SFI's 'Genius and Madness' Event in August," Santa Fe Institute, November 9, 2015, https://www.santafe.edu/news-center/news/genius-and-madness-sz-trans-english. Translated and reprinted with permission from Ulrike Duhm's reporting on "Genius and Madness" in the German daily *Sueddeutsche Zeitung*.

69   Bennis and Biederman, *Organizing Genius*, 17.

70   Woodward, "Cormac Country."

71   Thomas Halliday, "Portrait of the Human as a Young Hominin," *Nautilus*, May 25, 2022, https://nautil.us/portrait-of-the-human-as-a-young-hominin-18441/.

72 Romeo, "Cormac McCarthy on the Santa Fe Institute's Brainy Halls."

73 *The Unbelievers,* directed by Gus Holwerda (Hot Docs, 2013), DVD.

74 *Digital Transformation: Interview with David Krakauer.*

75 "The Quantum Field Theory on Which the Everyday World Supervenes," Santa Fe Institute, January 14, 2022, https://www.santafe.edu/news-center/news/quantum-field-theory-which-everyday-world-supervenes. Reprinted from *Santa Fe Magazine* no. 2, January 2022.

76 John Jurgensen, "Hollywood's Favorite Cowboy," *Wall Street Journal,* November 13, 2009, https://www.proquest.com/newspapers/hollywoods-favorite-cowboy/docview/399070032/se-2?accountid=7098.

77 Mitchell M. Waldrop, *Complexity: The Emerging Science at the Edge of Order and Chaos* (New York: Simon & Schuster, 1993).

78 "Laurence Gonzales Named an SFI Miller Scholar."

79 "One Reporter's Account of SFI's 'Genius and Madness' Event in August."

80 Dottie Indyke, "Symposium Leaders Explore Simplicity, Complexity," *Santa Fe New Mexican,* December 4, 1998.

81 *Santa Fe New Mexican,* February 25, 2000, 60.

82 D. Quentin Miller and Josh Dwelle, "Cormac McCarthy," *Contemporary Novelists,* Cengage/Encyclopedia.com, April 25, 2022, https://www.encyclopedia.com/education/news-wires-white-papers-and-books/mccarthy-cormac.

83 "Santa Fe Institute," ProPublica Nonprofit Explorer, https://projects.propublica.org/nonprofits/display_990/850325494/2012_12_EO%2F85-0325494_990_201012.

84 Romeo, "Cormac McCarthy on the Santa Fe Institute's Brainy Halls."

85 Dowd, "The Santa Fe Institute," 34. Dowd explains how the concept of emergence relates to complexity through the truism that the sum is greater than the parts (thus the wetness of water "cannot be deduced from a close analysis of its molecules of two hydrogen atoms and one oxygen").

86 Dowd, "The Santa Fe Institute," 36–7.

87 Ibid., 33.

88 Cormac McCarthy, 1968–1989, undated.

89 Pp. 181–5.

90 Timothy Andersen, "Quantum Wittgenstein," *Aeon,* May 12, 2022, https://aeon.co/essays/how-wittgenstein-might-solve-both-philosophy-and-quantum-physics.

91 "The Quantum Field Theory on Which the Everyday World Supervenes."

92 *Digital Transformation: Interview with David Krakauer.*

93 Andersen, "Quantum Wittgenstein."

94 "One Reporter's Account of SFI's 'Genius and Madness' Event in August."

2

Science

Starting from a Unified Place—How Chirality and Handedness Inform McCarthy's Expanding Universe

It begins in the untended riot of vegetation in Knoxville. In nature's time-so-deep-as-to-be-timeless, the ceaseless activity of evolution (McCarthy calls it "delicate cellular warfare in a waterdrop" [13]), the longest striving of all that presents itself, paradoxically, in the guise of unexamined stillness and resilience at the margins of violent human incursions. It starts with an innocuous detail in the opening passage of *Suttree* (1979) that observes the way that certain vines wind and spiral. In *Suttree*'s "gray vines coiled leftward" (3) a world begins unfurling. Its tendrils reach into the construction of all material things, so far as can be known, from the atoms that comprise us, to the unusual biology of identical twins, to the question of what constitutes an aberration in the fabric of the universe, and the relation between material and moral substance. So much can depend on an observed detail.

At its simplest, chirality is the property of the handedness of things, and it must be conceded at the outset that handedness is itself a conceit deeply rooted in the firmament of human experience and our very way of seeing the world—one so intrinsic to our worldview, in fact, that we rarely think about it. Anthropomorphically speaking, we can refer to objects ranging from spirals, wood screws, and plants as left-handed or right-handed based on the way that we apply our conveniently opposable-thumbed paws to the turning of them. Although it is none too intuitive, being able to distinguish

FIGURES 2.1 AND 2.2 *Chiral spiral petroglyphs on boulders at bottom of toe slope, Petroglyph National Monument, Albuquerque, New Mexico. Around 90 percent of the park's petroglyphs are attributed to by the ancestors of today's Pueblo people, with the majority etched from the 1300s to the 1680s, according to the US National Park Service. "What are Petroglyphs," US National Park Service, Petroglyph National Monument, last modified March 20, 2021, https://www.nps.gov/petr/learn/historyculture/what.htm. Photographed in 2022 by the author.*

between left and right is counted a developmental milestone even as it imposes a kind of semantically inscripted prejudice in our thinking and way of seeing (consider the mischief worked on the Western mind in those earliest dualistic lessons of childhood: *inside/outside, clean/dirty*, and the very important *left/right*).

There is nothing intuitive about the left/right distinction, as anyone who has taught a child knows. My father, a veteran of U.S. Marine Corps, often recalls how he was asked to teach left and right to certain recruits who had missed this lesson, a distinction that takes on a certain compelling urgency in a fields-of-fire situation. It is also innately referential and subjective: Martin Gardner posited the "Ozma Problem," noting that it would be impossible, without looking at the same object, to communicate the distinction between left and right if ever we communicated with an extraterrestrial intelligence. Previously, Immanuel Kant and William James had wrestled with dimensions of the same issue. Finally, the 1956 "Wu experiment," based on the beta decay of cobalt-60, disproved to physicists' satisfaction the conservation of parity, thereby making it possible to describe an experiment that could be replicated to definitively convey the meaning of left and right.[1]

Of course, explaining the difference between left and right via nonconservation of parity in beta decay is a bit like building a refrigerator to explain the concept of cold, and points to the strange difficulty underpinning our conventional description of direction. Taken for granted, the everydayness of our left/right framework for seeing the world obscures many fundamental mysteries: How do we know the difference between left and right without reference to additional planes? How does the human body, in its very construction, know the difference, and locate certain organs according to a left-right bias? Why are some things (indeed, most living things) more right-handed?

To have a look at chirality and how it operates in the imagination of Cormac McCarthy we must begin by dispensing some scientific terminology that may not be familiar to literary scholars. And it will be important to ask some forbearance on the part of the nonspecialized reader with the assurance in advance that concepts as various and apparently unrelated as physical cosmology and the handedness of plants (for plants can indeed be classed as left- and right-handed) will find illustration in McCarthy's literary work. The first half of this chapter looks at the science of chirality, and the second half at how it is manifested primarily in one of McCarthy's novels.

The organizational principle for this chapter is, fittingly, one of mirror symmetry: the second half really cannot be well understood without the first. And a reader who is interrupted at frequent intervals with literary illustrations might have a difficult time following the thread of the scientific context, which will be provided here with broad strokes and a certain amount of necessarily gross simplification for the sake of reaching the literary analysis in the second section.

A vast scientific literature of chirality awaits the curious, though much of it is highly specialized. The most important book on the topic is remarkably accessible, however. Oklahoman and popular science writer Martin Gardner's *The Ambidextrous Universe*, first published in 1964, was retitled *The New Ambidextrous Universe* and went into a third revised edition (1995) to include later developments such as string theory. Gardner's indispensable book is unlikely to be surpassed for its playful but sophisticated insouciance in laying out the many dimensions of handedness, symmetry, and asymmetry in the world. In fact, it is credited as an influence on science-minded Vladimir Nabokov, whose work Gardner cites in his own. Gardner would in turn be cited by Nabokov in *Ada, or Ardor: A Family Chronicle*. In fact, Gardner's hunch that Nabokov's *Look at the Harlequins!* (1974) had been influenced by his study of symmetry and asymmetry was later confirmed by "two scholarly papers—one by a literary critic, the other by a historian of science."[2] It is entirely possible that McCarthy took some influence from Gardner's work as well, and later in this chapter I will document the influence on McCarthy's work of at least one author who wrote about chirality.

For purposes of this chapter, which focuses on the biology of chirality, James P. Riehl's book *Mirror-Image Asymmetry: An Introduction to the Origin and Consequences of Chirality* (2010) offers a layperson-friendly overview of the topic.[3] For understanding how developmental anomalies are related to mirror-image asymmetry, chirality, and other factors, my colleague Lewis I. Held Jr's latest book, *Animal Anomalies: What Abnormal*

Anatomies Reveal about Normal Development (2022), is both authoritative and delightful (chapters are structured as Sherlock Holmes–style mysteries, to be solved by the reader).[4] With the determined sleuthing of an Arthur Conan Doyle, Held and his coauthor, Stanley K. Sessions, recently solved a 125-year-old riddle of biological symmetry, Bateson's rule. They revealed the reason why extra legs branching from the same circuit tend to be mirror-symmetric (spoiler alert: "it is an emergent property of the circuitry of the pathways and their polarized alignments along the limb axes").[5]

McCarthy explores chirality in several of his works—consider, for example, the overtly chiral pairing of the characters Black and White in *The Sunset Limited* (2006)—but the majority of the examples offered here will come from *Suttree*, since it is more concerned with the biological in its exploration of symmetry and asymmetry. And an understanding of chirality should perhaps be grounded in biology and our chemical composition, as these things, too, are reflected in the text.[6]

Chirality offers some lessons about how we indeed spin like planets, and how mirror-image opposites, when reconciled, speak to the importance of asymmetry, of being threaded through oneself as a means to spiritual revelation. Make no mistake: the best evidence suggests that the universe, broadly speaking, is asymmetric, though we might prefer, as a cognitive matter, to revert to the simple dualisms of chirality. Ethan Siegel writes, "During the 20th century, the recognition of certain symmetries in nature led to many theoretical and experimental breakthroughs in fundamental physics. However, the attempt to impose additional symmetries, while theoretically fascinating, led to an enormous series of predictions that weren't borne out by experiment or observation."[7]

"Today," he acknowledges, "many claim that theoretical physics has stagnated, as it's clung to those unsupported ideas. We must face reality: the Universe is not symmetric." So to challenge dualistic thinking at the outset we might recognize that (asymmetrically) there is more than one symmetry! For example, poet/scientist Katherine Larson's *Radial Symmetry* (2011) takes a close look at aquatic cnidarians—think jellyfish, sea anemones, corals, comb jellies and their five-pointed cousins, certain starfish, sand dollars, and sea urchins—which have a symmetry different to our own.

As we wonder at our universe (or multiverse, as the case may be), we might hope to get beyond the suppositions of bilateral symmetry that would pair science and poetry as "opposites," and go back to the sources of things, a single center, the unified place. Polish poet Anna Kamieńska writes of how it is

> To be transformed
> To turn yourself inside out like a glove
> To spin like a planet
> To thread yourself through yourself

So that each day penetrates each night
So that each word runs to the other side of truth.[8]

Chirality and the interplay of symmetry and asymmetry is a transformative concept in complex natural systems, in human experience down to embryonic development, and in the philosophies of deep-seeing writers. Understanding chirality in turn reveals new dimensions of scientific and philosophical meaning in McCarthy's work. What I am proposing is a new way of reading his work, and for some, it may offer a new way of seeing the world around us. I cannot pretend to answer all the questions I raise here, but neither can the current science surrounding the issues. My looping is instead a design, echoing McCarthy's, in the hope that you might *be drawn through yourself and run to the other side of truth*.

* * *

Before applying chirality to McCarthy's work, this section provides a brief overview of its scientific understanding. First, the term "chirality" derives from the Greek word for hand, and it is one of many coinages from Lord Kelvin, whose fertile mind was perhaps better suited to the twentieth century than the nineteenth.[9] In simplest terms, a system is chiral if the mirror image of the system cannot be superimposed on the original system. The most familiar example is at arm's length: it would be very odd if you were born with two left hands, in which case no one could accuse you of twiddling your thumbs! In simplest terms, then, chirality is the property of handedness. Some things that partake of chirality are the same except for being reversed along one axis (our mirrored hands, for example), thus the common phrase, mirror-image asymmetry—a term of art relevant to *Suttree* and to which we shall return.

Chirality turns out to be profoundly consequential at the molecular level. The molecules in the cells of living things are chiral, and, as James Riehl notes, "For more than 100 years scientists have known that the building blocks of proteins and DNA are composed of only one of the two mirror-image forms of amino acids" and a "special group of [right-forming] sugars."[10] Thus "exclusive homochirality would seem to be a unique aspect of life (at least life as we know it!),"[11] and it remains to be seen if the universe offers exceptions to our terrestrial rule. All naturally occurring proteins from all living organisms consist of amino acids with left chirality; it was established by the mid-twentieth century that sugars forming the helical structure of DNA and RNA have right chirality. Think of these, if you like, as the Legos by which living things are made, with a simple interlocking pattern to ensure stackable compatibility. On the other hand, consider the function of incompatibility to prevent miscoding, not unlike the way that a piece of Ikea furniture is engineered so that a wrongly inserted part will not connect with a mismatched one.

DNA is right-handed, in that the helix spirals in a clockwise direction; there are websites devoted to exposing, with nerdish glee, scientific bloopers in media illustrations depicting a DNA molecule quite impossibly turning in the other direction. Biological molecules are by nature chiral because they are built from single enantiomers (each one of a pair of chiral molecules that mirror each other). And amino acids, the essential ingredient of living things, furnish chiral building block molecules. It is precisely because biology is chiral that the chirality of drugs turns out to matter quite a lot, as different enantiomers can produce very different reactions to the body's chiral architecture. The classic case of this is thalidomide, which was widely prescribed as a sedative for pregnant women during the 1970s. One of its enantiomers makes an effective sedative, and the other, tragically, causes severe birth defects. Moreover, the drug as processed (interconverted) in the body makes a racemic mixture— one in which left- and right-handed enantiomers of a chiral molecule are found in equal proportion—so the benign enantiomer cannot be guaranteed.

The thalidomide problem has an American pop culture cameo when Walter White lectures on its textbook example in *Breaking Bad* (Season 1, Episode 2); he will later bargain for his life by pointing out that his methamphetamine is "enantomerically pure" (Season 4, Episode 1), adding some additional currency to the mirror-image motif of the series, in which reflected images and characters (like White-Fring) abound. As *Vice* journalist Jason Wallach points out, White might have used the example of methamphetamines for chirality: "d-methamphetamine induces classic stimulant effects, whereas l-methamphetamine is only a weak stimulant but an excellent decongestant, which is sold over-the-counter in Vicks® inhalers under the pseudonym les-desoxyephedrine."[12] Needless to say, any chiral drug marketed as a racemic mixture must now be proven pharmacologically and toxicologically safe for each enantiomer. When one enantiomer is as effective as another, American pharmaceutical companies have gamed the system by "flipping" a molecule and securing a fresh patent just before intellectual property rights expire and a drug becomes generic.

In any case, chirality is so intrinsic to organic chemistry that entire journals are dedicated to its study, including, naturally, *Chirality*, which celebrated the award of a 2016 Nobel Prize in Chemistry to longtime contributor Ben L. Feringa. An important branch of evolutionary biology is devoted to sorting out the presentation and significance of chirality in living things. Its study entrains a range of fields of scientific inquiry, including physical cosmology. In fact, an important goal of the Comet 67P/Churyumov-Gerasimenko spacecraft landing was to see if chiral molecules exist in space (the study of meteorites is mostly inconclusive because of the likelihood of tainted samples). The discovery of chiral molecules there would lend some credence to the comet "fertilization" theory—that a comet delivered some of the complex organic molecules and L-amino acids that set earthly evolution in motion.[13]

Other theories abound as to the origins of chirality on earth, including the influence of the nuclear non-parity-conserving weak interaction and even the way that neutron stars generate UV circular polarization. Indeed, atoms and their particles themselves demonstrate chirality in their spinning, so it may fairly be said that chirality informs the very smallest things revealed to us by scientific observation. It also begs a reconsideration of what, precisely, we mean when we speak of chiral molecules, since they consist of atoms with chirality, fermions with chirality, and even massless particles. If racemic or "opposite" chirality (for instance, a preponderance of r-amino acids) were discovered elsewhere in the universe it would suggest the part of chance in determining chirality here on earth going back to the very beginnings of things. Writing in *The Guardian*, Andrew Rutherford was prepared to say, "A mirror version of DNA could perfectly well exist, but it appears that the coin was flipped about 4bn years ago, and every life form since has twisted right."[14] This conclusion, however, seems at best premature and at worst a bit anthropocentric, at least until we are acquainted with denizens of more distant reaches of the cosmos.

For now, suffice it to say that we may be having something of a chirality moment, from *Breaking Bad* to Nobel Prize–winning areas of scientific inquiry. Recent confirmation of assymetry in the arrangement of galaxies has prompted speculation that another Nobel prize might be in the offing in this area. Chirality obtains both in the origins of the very smallest things and in the larger forms of living things, and it is the latter that we turn to now. Organisms themselves can be symmetric, racemic (occurring in equal proportions of left-handed and right-handed individuals), or asymmetric. In the first category, animal body plans take a variety of symmetries, including radial, chiral, bilateral, and spherical. The human plan, as it happens, is pseudo-bilateral: we like to think we're nicely symmetrical along the vertical (dorsal ventral) axis. Imagine a paper cutout of the human form, with dotted line from the nose to the navel; fold, and the two sides would correspond neatly through mirror-image symmetry.

In reality, of course, the symmetry is far from perfect (take an honest look at yourself and notice your two differently sized feet, breasts, testes, etc.), though studies of the perception of human faces show that we are hard-wired to see the symmetrical as the beautiful. Beauty, in that sense, is truly skin-deep. When Nick Romeo asked Cormac McCarthy if "a beautiful sentence [was] more likely to be true in some way," McCarthy chuckled and responded with a short lesson in aesthetics that evoked, naturally, symmetry:

> That's tough. It's hard to define beauty, though there've been some strange attempts. We know it involves harmony, repetition, symmetry. These things speak to us and have for a long time.[15]

Our physical construction relies on harmony and repetition—the scaling up of simple biological formulas—but one need only look inside to see that we are thoroughly asymmetrical, beginning with such familiar example as the left-sided heart and the positioning of the organs within the body. In summary, "Vertebrates have a generally bilaterally symmetrical body-plan, but this symmetry is broken by the consistently asymmetric placement of various internal organs such as the heart, liver, spleen, and gut, or the asymmetric development of paired organs (such as brain hemispheres and lungs)."[16] There is chirality in the corkscrewing of your intestines, for instance. And this anatomical chirality has important consequences that are only beginning to be understood: "The mechanisms which ensure invariant LR asymmetry of the heart, viscera, and brain represent a thread connecting biomolecular chirality to human cognition," writes Levin, "along the way involving fundamental aspects of cell biology, biophysics, and evolutionary biology. An understanding of LR asymmetry is important not only for basic science, but also for the biomedicine of a wide range of birth defects and human genetic syndromes."[17]

Before looking at some of those consequences of asymmetry, however, consider chirality as it obtains broadly to the natural world around us. Chirality might be a familiar concept in terms of our species, but it prevails among many other organisms, including, for example, insects that have exquisitely evolved communication systems based on chiral pheromone signatures. And then there are plants. Consider a sprouting vine that reaches a horizontal filament of wire. Will the plant wind around the wire to the left or the right? It turns out that roughly 90 percent of the world's plants favor a right-handed helix, which is to say (somewhat confusingly) that they wind in an anticlockwise fashion from the ground up. Contrary to common conception, plant twining does not seem to be correlated in any way to the sun's movement through a particular celestial sphere, by the Coriolis effect, or by any other hemispheric influence.[18] The other roughly 10 percent of plants are either lefty-contrarians or unfussy-ambidextrous sorts.

Curiously, the percentage of right-handed plants is comparable to the global average of right-handedness observed in people. Handedness in humans is a non-heritable, non-Mendellian trait, so it cannot be predicted in the classic dominant-recessive way. Handedness varies among identical twins at about the same rate as handedness in the general population.[19] Which suggests that the interaction of genes with environment likely drives handedness, implicating the process of socialization—and perhaps the notably commonplace practice of training children to write with their right hand. Right-handed chimpanzees are twice as common as left-handed chimps, and some claims have been made for the increasing occurrence of right-handedness in human populations, based on sources ranging from cave painting to the analysis of ancient remains.

In Levin's summary, "A huge literature on brain lateralization phenomena in human beings exists as well, but many of these asymmetries are secondary and arise as a result of cultural environmental biasing factors."[20] Which is another way of saying, despite the outpouring of literature about left- and right-brained people and handedness crossover, much of what you think you know about the topic may be wrong, and despite many theories, from spear throwing to socialization, no one knows exactly why human right-handedness prevails globally and across cultures.[21] Add to this a good bit of web-based mythology, including, for example, the widespread (and debunked) notion that polar bears are predominantly left-handed, and it soon becomes apparent that countless studies have done little to clarify the essential facts of southpaws in any species.

Having said that, some interesting and legitimate studies have been done on predator–prey interactions and handedness. For example, the overwhelming majority of snails are right-handed (their shells turn clockwise). Various reasons have been postulated, including the alignment of cellular microtubules, but another line of explanation proceeds from the idea that their chirality could confer a selective advantage depending on the handedness of predators, such as snakes or crabs. The limbering of a sea snake's jaws and its angle of attack, for example, might make one snail's chirality easier to swallow than another. So, there are indeed cases where selective pressure can explain handedness. Interestingly, mutations causing a switch in handedness are correlated with speciation events, since, for example, opposite-handed snails have mechanical trouble mating. Some species, like the familiar fiddler crab, are racemic in their handedness; if you see a group of them, you might notice that the dominant claw occurs roughly equally on the left and the right. There are also well-documented cases, such as lobster claw morphology, in which chirality is determined by neurological activity. Spending one's whole existence reaching out from a craggy niche in the same direction would suggest that it is best to play to the circumstances.

Taking a crude look at the whole, biologist Sarah Huber explains,

> There are several lines of evidence to suggest that "handedness" is often selectively neutral and simply the product of chance. First, we see both left- and right-handed organisms, and these organisms have overlapping distributions and ecological niches. Second, handedness seems to "run in the family." Organisms that are closely related and share an evolutionary history have the same direction of spin—what biologists call phylogenetic inertia. A trait evolves in a lineage, and over time it persists because there are no forces to change it. Handedness is a homology, shared because of common descent.[22]

So, we come full circle, with handedness as a coin-toss, absent some force to change the bias. Notwithstanding the sophistication of the scientific understanding of molecular chirality and genetics, it is astonishing how questions of handedness continue to bring us back to sensitive dependence on initial conditions, aka complexity. Levin writes,

The establishment of left-right (LR) asymmetry raises a number of fascinating biological questions. Why does asymmetry exist at all? What are the implications of asymmetry for the normal structure and physiology of the heart, gut, and brain? Why are all normal individuals not only asymmetric, but asymmetric to the same direction (i.e., why a consistent bias and not a 50/50% racemic population, given that individuals with full inversion are not phenotypically impaired)? When, during evolution, did handed asymmetry appear, and were there true bilaterally symmetrical organisms prior to the invention of oriented asymmetry? Is it connected to chirality in lower forms (such as snail shell coiling and chirality in some plants)? At what developmental stages is asymmetry initiated in vertebrate embryos? How conserved are the molecular mechanisms establishing correct asymmetry in animals with drastically different modes of gastrulation? And, how can the LR axis be consistently oriented with respect to the anterior–posterior (AP) and dorso-ventral (DV) axes in the absence of any macroscopic feature of chemistry or physics which distinguishes left from right? Answers to these questions require a detailed understanding, at the molecular, genetic, and biochemical levels, of the formation of biased asymmetry in embryos.[23]

How can nature establish the template for left-right (LR) asymmetry, and orient cells on that axis: it is one thing for the heart to be on the left side of the body, but how in the world do the body and the cellular material that comprises it know which side is the left? By the fourth or fifth week of pregnancy, cells receive the instructions to determine the laterality of organs—as Levin and his coauthors waggishly titled a paper concerning the role for the cytoskeleton in LR asymmetry, "It's never too early to get it Right." Recently the contest between the two major theories of interpretation—one, that L-R asymmetry is primarily set up by nodal cilia, and another, by ion pumps—has tipped toward the nodal cilia model for most vertebrates, including humans. One clue is that ciliary dyskinesia syndrome causes situs inversus (flipping of the organs positions in humans—more on this later).

Still, if natural selection can be described as frozen accidents and the "non-random survival of random variants" as Richard Dawkins is sometimes paraphrased, what evolutionary reason is there for chirality? Incredibly, "explanted linear heart tubes from chicks or fish develop dextral looping in culture, indicating that this morphogenesis is independent of the LR body axis."[24] In other words, organs can develop LR asymmetry even when they have no body to provide a frame of reference (thank you, nodal cilia)!

As with so much in biology that seems merely stochastic, there is the temptation to discern a pattern and call it order. Yet even in cases where genetic information has been scrambled, there are observable patterns. For example, in human hermaphrodites, ovaries consistently develop on the left, testes on the right. Laterality defects in conjoined twins almost always afflict the twin on the right. More recent, identical twins have been observed in which only one has the symptoms of the Zika virus, leading to the theories regarding how the virus might cross the placenta[25]—only one of many examples where understanding the role of mirror-image asymmetry is important to preventing birth defects.

Moreover, twins, particularly identical twins, sometimes demonstrate mirror-image asymmetries. And "non-conjoined monozygotic [identical] twins, while not exhibiting the kinds of visceral laterality defects that occur in conjoined twins, do manifest many subtler kinds of mirror-image asymmetry ('bookend' or enantiomer twin pairs)," notes Levin. Citing diverse studies, he observes, "Pairs of such twins have been noted to present mirror asymmetries in hand preference, hair whorl direction, tooth patterns, unilateral eye and ear defects, cleft lip, cleft palate, supernumerary teeth, and even tumor locations and undescended testicles."[26] Some researchers have suggested that "some as yet unknown pathological mechanism is responsible for both the process of twinning itself and the destabilization of the LR axis";[27] similarly, it has been speculated that a small number of left-handers are "pathological" left-handers, meaning that their chirality is caused by an abnormality.[28]

All of these will be relevant to the discussion of handedness in *Suttree* in the second half of this chapter. For now, suffice it to say that we inhabit a living world of handedness, of spinnings, leanings, and spiralings, all of which are definitional in evolution. Twins illustrate the action of chirality in the flesh. More broadly, though, chirality turns out to be critical in the way that life forms, is sustained, and changes. And if this holds true to the earliest stages of our biological formation and evolution, we should not be surprised to find it reflected in literature and philosophy as well.

* * *

But if you ride these monsters deeper down, if you drop with them farther over the world's rim, you find what our sciences cannot locate or name, the substrate, the ocean or matrix of ether which buoys the rest, which gives goodness its power for good, and evil its power for evil, the unified field: our complex and inexplicable caring for each other, and for our life together here. This is given. It is not learned.

—ANNIE DILLARD, *TEACHING A STONE TO TALK* (2009)[29]

On the opening page of *Suttree*, McCarthy sets his tale in deep time and the fossil record, accurately describing the stones mined from the Cumberland plateau that contain "in their striae fossil bones, limestone scarabs rucked in the floor of this once inland sea," (3) the latter clause accurately reflecting the presence of marine organisms in the fossil record dating back to the time when the entire area was submerged and part of the sea floor. There is an evolutionary order to the opening as the reader soon reaches, in the same long passage, "Gray vines coiled leftward in this northern hemisphere, what winds them shapes the dogwhelk's shell" (3). What winds them? One answer, naturally, is chirality: the plants are wound by chirality, they exhibit chirality, because they are wound. This is a circularity; one may as well say, "Because nature made them so," but the more we understand of the processes of chirality, the more deeply we peer into the mysteries—or, in SFI parlance, the complex systems—of the natural world. McCarthy is pointing to the handedness of vines, and, as was discussed earlier, the ones described here coil leftward, placing them in the majority. Either Homer nods, or can be forgiven for subsequent discovery, because McCarthy seems to give some attribution to the now-disproven theory of hemispheric influence on plant-winding: what winds them is not the Coriolis effect. Rather, what winds them is none too straightforward. One might point as well to the Fibonacci series as a repeated template in nature, or to a measure of evolutionary determinism in the common pattern of things represented in the coiled vines and dogwhelk shell. Both are classic illustrations of chirality.

FIGURE 2.3 *This chiral fossil is between twelve and thirteen million years old and was embedded near the city of Merida in the Yucatan Peninsula of Mexico—rucked among the striae from the seafloor. From the Gran Museo del Mundo Maya in Merida, Mexico. Photographed by the author, April 2017.*

FIGURE 2.4 *"What winds its shell?" Bisected fossilized nautilus shell exhibiting dextrorotary chirality (as well as the Fibonacci series). From the Gran Museo del Mundo Maya in Merida, Mexico. Photographed by the author, April 2017.*

McCarthy's interest in chirality finds clear confirmation in his notes. In box 19, folder 14 of the McCarthy papers in Wittliff Collections at San Marcos, his jots for *Suttree* contain a page labeled "Dexter-Sinister-Coriolis."[30] Michael Crews's painstaking master-study of McCarthy's sources, *Books Are Made of Books*, points out that at the center of those notes is Joseph Wood Krutch, "a literary critic and nature writer" whose critique of modernity revolves around humankind's distance from nature. One of McCarthy's notes reads, "See The Desert Year Krutch—Carlsbad bats." In *The Desert Year* (1952), Krutch observes that bats emerging from Carlsbad Caverns spiral in a counterclockwise pattern which he then attributes to the Coriolis effect. Fascinated by the tendency toward patterned self-organization in the natural world, Krutch was concerned about mechanistic determinism: "On the other hand, since I am rather prone to hope that there is not a mechanical explanation for practically everything, it would also be gratifying to learn that there was positively nothing at all in my Coriolis Effect theory. One would then be left to wonder just how the Carlsbad community came to agree upon its traffic laws."[31]

Indeed, this line of speculation goes directly to chirality and questions about the underpinnings of order in nature. As Crews affirms, "McCarthy's interest in Krutch's reflections find expression in his use of symmetry as a recurring motif in Suttree." The notes on "Dexter—Sinister—Coriolis" give various textbook examples of chirality in nature: "Plaice flounder sole in the tropics are lefteyed . . . while those in the north are dextral. But the halibut is sinistral in both hemispheres—perhaps just for the halibut."

There are anthropological examples, too: "Hottentots and bushmen 70% sinistral. Their austral nature." This amounts to a leap of logic, and the influence of hemispheric influence has largely been scientifically discredited, as mentioned earlier. But it nevertheless finds its way into *Suttree*, with "austral" appearing, as Crews notes, in Suttree's remembrance of his Uncle Milo, lost off the Chilean coast: "Foreign stars in the nights down there. A whole new astronomy. Mensa, Musca, the Chameleon. Austral constellation nigh unknown to north folk" (128). Moreover, Crews correctly notes that Suttree's entire sense of self is grounded in the notion of the right/left and the sinister. In the same "Dexter" notes McCarthy has copied a quotation for Philip Massinger's play, *The Virgin-Martyr*, "Left-ey'd knight of the antipodes Philip Massinger."[32]

In fact, the very first pages of *Suttree* announce that it will also be in some sense a study in mirror-image chirality and a/symmetry. The "murengers" have walled in a presence ("the thing's inside" (4)) and the question is posed, "[W]hat's the counter of his face?" Shortly after the movement of a cat across cobblestones is described as "sewn in rapid antipodes over the raindark street to vanish cat and countercat" (4) in the first of numerous doublings in the novel.

Indeed, the earliest pages of the novel find Suttree contemplating, in stream-of-consciousness, "The delicate cellular warfare in a waterdrop. A dextrocardiac, said the smiling doctor. Your heart's in the right place" (13). From the standpoint of chirality, there isn't much that's amusing about dextrocardia; having the heart on the right side guarantees a short life, although complete mirror reversal of internal organs (situs inversus) is generally harmless and, as mentioned earlier, might vindicate the school of thought that posits nodal cilia as the source of LR symmetry in humans. The "warfare in a waterdrop" is evocative of the zygote and cellular conflict described 100 pages later. Suttree observes the lamplight and shadowy corners "[w]here insect shadows war. The reflection of the lamp's glass chimney like a quaking egg, the zygote dividing. Giant spores addorsed and severing. Yawing toward separate destinies in their blind molecular schism. If a cell can be lefthanded may it not have a will? And a gauche will?" (113). The image is important enough that he will return to it, self-referentially, much later ("You can tell me, paradigm of your own sinister genesis construed by a flame in a glass bell" [414]), and this is arguably one of the most pregnant passages in *Suttree*.

It touches on Cornelius Suttree's stillborn identical twin even as it raises a question about the nature of evil, determinism, and free will. There are ways in which it might be said that nature chooses sides—though it should be conceded this language of agency, of a cell having a "will," which McCarthy employs in *Suttree*, is nothing more than a convenient conceit. Did the universe will a preponderance of L-amino acids? Or was this simply a matter of chance, a cosmic coin toss of the kind that fascinates McCarthy?

Note that in *No Country for Old Men* (2005) Chigurh couches the outcome of the ultimate coin toss in terms that are suggestive of chirality: "Every moment in your life is a turning and every one a choosing. Somewhere you made a choice. All followed to this. The accounting is scrupulous. The shape is drawn" (152).

Returning to the fundamental imponderability of the origins of chiral asymmetry—genes, environment, or chance—it might be said that all become determinisms. A turn, a bend in the universe, and the shape is drawn. As mentioned earlier, one line of thinking is that the die was cast for chirality in the earliest moments of the universe. Perhaps the early arrangement of the universe set the stage to create UV circular polarization from white dwarfs, neutron stars, and so on. Perhaps the course was set by the fact that "in 10^{18} molecules (a million trillion molecules) there would be one more L-amino acid than D-amino acid due to [the] nuclear nonparity-conserving weak interaction."[33] With world enough and time, even the slightest bias could tip the scale in complex systems.

One might easily go to an ontology here. Is it possible that there is a rightness to things, since the universe we inhabit so favors this asymmetry? And if that rightness is a kind of inscription ("given, not learned" as Dillard writes), does it furnish a constant in the equations of natural law, or is it a law in its own right? Much has been made of the moral valence of "right" and "left," so it's a rather striking coincidence that our part of the physical universe indeed seems predisposed toward the rightness of things. Put differently, from our limited observations, the universal script slants to the right. Elsewhere I have written about the importance of the concept of logos in McCarthy's work—and logos is the ordering principle of the Christian universe (e.g., "In the beginning was the Logos"). The question I wish to pose here is whether chirality in some way underpins the logos, as an essential element of whatever natural and by extension moral order there be, in contrast to what McCarthy terms the *gauche will* in *Suttree*.

Before moving along, that leap of logic might best be spelled out this way:

1) Chirality imposes an observable bias throughout the universe, distinct from what we might expect to witness (a racemic, 50–50 universe) if pure chance controlled.[34]

2) There is an ordering principle at work in the universe.

3) This is evidence of design, since design alone can account for a favored direction throughout nature.

And it is indeed a leap of logic, for the third statement completes the circularity. Even if the prevalence of chirality contests randomness on one level, it is itself no refutation of material determinism, especially since we can only speculate about the chain of causation, and our observations are very limited (consider chaos theory's emphasis on sensitive dependence on initial conditions). Moreover, many would say that order does not of itself require a designer; rather, order simply *is,* or is created through the circularity of its conformity with our sense-inferring rules describing it. A good deal of order has already been inferred in random appearing/chaotic phenomena through chaos theory and differential equations. So it might be said that randomness determined the original tipping, after which the bias became general. Perhaps it makes the existence of opposite chirality—call it the sinister aspect—all the more intriguing. Is the cell that rejects the dominant chirality in some sense, as McCarthy suggests in *Suttree,* "willful"?

The material determinism to which McCarty returns relentlessly throughout his opus raises questions of perception, narrative, and witness as well. Could it be that the true metaphysic is within our very substance? Consider this syllogism and its relationship to the last example:

1) The chiral nature of the physical universe inscribes a quality of handedness in the material.

2) Our very substance, then, records the direction of the origination of all things.

3) Ergo it would seem that metaphysical reality is written within the physical world and within us, and that this understanding, even when ineffable, is nonetheless with us.

This conclusion chimes with David Tracy's *The Analogical Imagination* and a long tradition of theologians countenancing a creator who is more like the created world than might be expected.[35] Compare this line of reasoning to this passage describing the heretic's realization in *The Crossing,* which I've parsed as an "argument" here:

1) He saw that his demands upon God resided intact and unspoken also in even the simplest heart. His contention. His argument. They had their being in the humblest history.

2) For the path of the world also is one and not many and there is no alter course in any least part of it for that course is given by God and contains all consequence in the way of its going and outside of that going there is neither path nor consequence nor anything at all. There never was.

3) In the end what the priest came to believe was that the truth may often be carried about by those who themselves remain all unaware of it. They bear that which has weight and substance and yet for them has no name whereby it may be evoked or called forth. (157–58, my line breaks)

McCarthy enters into the territory of apologetics here, reflecting many traditions, including, I think, an Aristotelian predilection for a prime mover, though elsewhere he plays on Platonic forms. There is little reason to assume that these views are McCarthy's own, or to accord the final say to the priest here—at least no more so than any other of McCarthy's characters who speak their truth. There is, however, reason to give some authoritative weight to the notion of the teller and the tale. It is generally conceded that the Border Trilogy is a frame tale—consider the hushaby of the pretty horses lullaby that begins it, and the dream of Billy Parham that concludes it—conferring ultimate reality only to tale itself, right down to its final epigraph (". . . The story's told/Turn the page" [n.p.]). The priest finally concludes that "the lesson of a life can never be its own. Only the witness has the power to take its measure. It is lived for the other only" (158). Whereas "God needs no witness" (158).

Another syllogism, then, drawing on McCarthy's Christian narratology:

1) The ability to take a lesson presupposes a larger context for meaning. Human lives require witness "to take their measure" and be validated ("There is another who will hear what you never spoke" (158)). By contrast, God needs no witness. ("The priest therefore saw what the anchorite could not. That God needs no witness. Neither to Himself nor against" (158)).

2) Synthesizing various declarations from this section of *The Crossing* and elsewhere in the Border Trilogy, tales need no witness to be true and there is only one metanarrative of truth, ultimately (consider the truth of folk tales and allegories alongside a postmodern novel).

3) Ergo, the truth of a story must proceed from a place beyond signifier and symbol. Paradoxically, stories speak to our need to be witnessed in order to confer meaning, even as their real meaning proceeds from, and conducts to, a source of meaning that requires no witness. To that extent, stories could be a means of witnessing the inscription of godhead to/in us, even as their spiritual truth is obfuscated by language and our need to be witnessed. We comprehend the meaning of a story only as far as "what we have made of God" (158).

A further "proof" might be added to this, since tales are stamped with an irrefutable order if they are to have meaning. Consider this observation:

1) Space is multidimensional and multidirectional.

2) Time, in both narrative and human understanding, is unidirectional; it is a syntax for what happens in the universe, and in narrative controls a morphology derived from the expected order of things (observation conditions expectations of causation).[36] "A form without a history has no power to perpetuate itself. What has no past can have no future" (*Cities of the Plain* 281).

3) Therefore, there is an ordering principle at work in stories that gestures toward a common point of origin and the inscription of a *logos*.

Stories, then, contain an underlying metaphysical reality; as with the chirality of things, they necessarily contain direction that is independent of what they say (expression), but their ability to say anything is dependent on that direction.[37]

Where that direction comes from is the mystery that we fail too often to see. The chirality of DNA defines the mode of expression that is called a phenotype, and so it is with time and narrative. So directionally time-bound are we that while it is possible to write a story beginning in media res or even in reverse (a textbook example being F. Scott Fitzgerald's "The Curious Case of Benjamin Button" [1922], a story with a protagonist who ages in reverse), the story ultimately is understood within the mind of the reader, and indeed can only be meaningful, through its linear rearrangement from start to finish. "You can't rewind the universe and run it over again," says one of the characters in a draft of *Whales and Men*, reflecting the arrow of time and also sensitive dependence on initial conditions.[38] Former SFI professor John Holland writes, "For both CPS (complex physical systems) and CAS (complex adaptive systems) there is a time dimension that must be incorporated, in which an element is generated by the grammar (the placement of a word in a sentence) is explicitly attached to that element (in subject-verb-object grammar, a verb is ordinarily placed *after* a subject noun). In a broader context, the order of generation of elements in the corpus plays the role of time."[39] Either way, the logic of time is implicit.

As Martin Gardner points out in *The New Ambidextrous Universe*, "It is easy to see a mirror-reversed world. . . . But seeing a time-reversed world poses difficulties." Gardner examines this topic in some detail, pointing out that modern physics countenances the possibility of not just antiparticles and antimatter but an antiworld where time and space might be reversed. Gardner's complicated analysis spans chapters 28–31 of the book, investigating the reconciliation of free will and determinism, among other things. In one thought experiment, he muses that "two worlds with opposite time arrows are analogous to two worlds that are mirror images of each other," and in such a world, intelligent beings would be "living 'forward' in their time"—but that communication between time reversed worlds would be "ruled out by logic."[40]

This is indeed the stuff of science- and other-fiction genres, and some physicists think that time can flow backward, or that it doesn't exist.[41] But even when authors toy with nonlinear chronology, magical realism, and so on, it makes no matter, because time imposes a forward-running syntax on the truth of a story, and the story is its own place. As the caretaker of the ruined church says in *The Crossing*, "The story on the other hand can never be lost from its place in the world for it is that place. . . . And like all corridos it ultimately told one story only, for there is only one to tell" (142–3).[42]

I am no expert in narratology, and my interpretations here are probably as crude as my philosophical conclusions. These "proofs" are not going to satisfy logicians or theologians; rather, they are a thought exercise to show the main drift of argument.[43] And yet it seems clear that there is something important in both the dialectical and analogical dimensions of McCarthy's imagination. Garry Wallace remembered a conversation in which McCarthy said he "thinks the mystical experience is a direct apprehension of reality, unmediated by symbol, and he ended with the thought that our inability to see spiritual truth is the greater mystery."[44] This recapitulates McCarthy's distrust of language, and also rephrases several notions above. Spiritual truth—a logos—flickers up from the unconscious, the story-as-story (not as language), the meaning in the dream of life.[45]

The best way I know to sum this up is: language is to consciousness what matter is to chirality. To be clear, my purpose in all this is not to enter apologetics, or to offer an argument from design, or to go so far as to claim that McCarthy is affirming a teleology in some holistic way in his opus. If the language/chirality analogy holds, it brings a consistent mode of interpretation to McCarthy's constant interrogation of natural law. For if chirality bears some relation on what has conventionally been called natural law, then natural law, through thesis and antithesis (which, in the chiral way, are not "opposites" in synthesis—see the concluding chapter), is unquestionably an enduring area of interest in McCarthy's work. A twist on natural law is encapsulated in the thinking of *Child of God*'s (1973) Lester Ballard, who cannot begin to comprehend how hawks mate, but remains confident in the knowledge "that all things fought" (168)—a conclusion not dissimilar to Hobbes's revision of natural law presuming the state of "war of every man against every man"—the well-known doctrine of *bellum omnium contra omnes*.[46] For Suttree this awareness extends even to the microscopic things in a drop of water. If natural law dictates the goodness of life, what if life is inclined to be, in some sense, so zealously protective of itself as to be viciously amoral or inclined to so-called natural evil?

McCarthy had little chance of avoiding the complications of evolution. The "old tattered barrister" whom Suttree encounters had "been chief counsel for Scopes, a friend of Darrow and Mencken and a lifelong friend of doomed defendants, causes lost, alone and friendless in a hundred courts" (367). The figure of John Randolph Neal from the famed 1925 Scopes Trial was familiar to McCarthy's father (see Chapter 4). As a boy, one of Cormac McCarthy's chief hobbies was collecting fossils and other specimens in the hills around Knoxville, so it does not go too far to say that evolution was in the water he drank, the land he walked, and the people he encountered.[47]

TABLE 2.1 *A Brief Exercise in Symmetric and Asymmetric Tensions a la Cormac McCarthy*[a]

Language	(Un)consciousness
Script	Inscription
Evolution-susceptible	Ur-language
Symbol	Logos/meaning
Syntax	Sacrament
Matter	Chirality
Symmetry	Asymmetry
Teleology	Looping/gyring
Neo-Darwinism	Natural Law/motion
Idiom	Truth
Learned	Given
Site	Nonsite
Witness-dependent	Witness-transcendent
Story	Dream
Imperfection	Originality
Entropy	Grail
Substance	Tale (singular narrative)
Twinning	Unified self
Irreversibility	The centered

[a] In terms of chirality, it is interesting to consider which of these terms belong on the left, and which, on the right. Many, it will be observed, do not submit easily to such alignments, and can be flipped. Perhaps the valences of these terms, taken as a whole, might reveal how the slant of things informs understanding.

What comfort is the natural law so carefully drilled into a young Cormac in his Knoxville parochial schools when the primary lessons of Darwinian evolution invite conclusions such as the oft-quoted last sentence of *Suttree*'s prelude: "Ruder forms survive" (5)?[48]

Indeed, the much-quoted phrase from *Suttree* is a truism: in the textbook example of complexity science, nature relentlessly recycles the simple formulas of embryology, without regard to seemingly cruel outcomes. Moreover, it is a direct reflection of the evolutionary discourses in which Charles Darwin, not Jean Baptiste Lamarck, proved to be correct. Indeed, Lamarck saw evolution as driving biological systems toward complexity and perfection; Darwin was perfectly at peace with the propagation of all things dull and nasty. In *The Descent of Man, and Selection in Relation to Sex* (1871), Darwin pondered the relationship of his evolutionary theory to ethics:

> The virtues which must be practiced, at least generally, by rude men, so that they may associate in a body, are those which are still recognized as the most important. But they are practiced almost exclusively in relation to the men of the same tribe; and their opposites are not regarded as crimes in relation to the men of other tribes.[49]

In other words, proscriptions of murder within my tribe may have little bearing on my willingness to murder within yours; in *Suttree* one indeed sees the minimal virtues that permit rudimentary association.

Anticipating the misapplication of his own theories, Darwin urged that civilized human society could achieve a state that transcended mere survival of the strongest, in which a meek person could contribute just as much as an aggressive one. He recognized the limits of his own theory in this respect, and he was careful to point out that his speculations on the evolutionary society were in a separate and distinct stream from evolutionary biology. The microbiological understanding of genetics that led to neo-Darwinism was not yet available after all. The interesting question here is whether McCarthy's declaration that "ruder forms survive" flirts with social Darwinism, affirms evolutionary theory, or goes against it.

His interest in these matters extends as far as debates regarding whether consciousness and language are in some sense evolutionary. In "The Kekulé Problem" (2017) essay, McCarthy posited a separate channel, after the fashion of Darwin and his ethics, for language: "There is no selection at work in the evolution of language because language is not a biological system and because there is only one of them. The ur-language of linguistic origin out of which all languages have evolved." Reflecting further on this, McCarthy admits,

> Influential persons will by now of course have smiled to themselves at the ill- concealed Lamarckianism lurking here. We might think to evade it by various strategies or redefinitions but probably without much success. Darwin of course was dismissive of the idea of inherited 'mutilations'— the issue of cutting off the tails of dogs for instance. But the inheritance of ideas remains something of a sticky issue. It is difficult to see them as anything other than acquired. How the unconscious goes about its work is not so much poorly understood as not understood at all.[50]

Thus the author confirms his interest in the divide between Lamarck and Darwin. It is not the only place where McCarthy's probes the limitations of Darwinism. As Michael Crews points out, McCarthy's dialogue fragments contain a note about Gordon Rattray Taylor's *Great Evolution Mystery* (1981), a critique of conventional Darwinism that McCarthy "seems to have taken seriously," in Crews's estimation.[51] "[Darwinism] dismisses the captain and puts the stoker at the helm," concludes one character in *Whales and Men*—an oddly teleological declaration if it is taken for McCarthy's view—in which the formula does not need its creator, and the unconscious does not need the conscious.[52]

McCarthy's interest in the relationship between semantics and consciousness has been a point of rumination going back at least to that unpublished screenplay, when Irish aristocrat Peter Gregory associates the formation of linguistic consciousness with the problem of evil, asking, "What was it that had made us outcast in this paradise created for us. What had made us refugees from joy and orphans of delight? What was it that characterized our species, that was found nowhere else in nature?" The answer, naturally, is language, which contains and supplants our experienced reality, leading Peter to the conclusion that it led to the very expulsion from Eden: "We were put into a garden and we turned it into a detention center."[53] Viewed this way, language containerizes reality, synthesizes and limits it in turn. Consciousness, as we experience it, is mediated by symbol (going back to the prelingual cave man epiphany, as McCarthy puts it in the Kekulé essay, that "one thing can be another thing") and expressed through the wholly inadequate, affected mirroring of language. The unconsciousness, stamped with our deepest animal origins, is the thing.

Thus, one of the essential properties of language in McCarthy's view is that it imbues our reality with a quality of secondary-ness. This notion is encapsulated by dialogue between Alicia and her psychiatrist in *Stella Maris*:

> But I think you've suggested that the advent of language, aside from the enormous value of it, was disruptive.
>
> Very disruptive. Of a piece with its value. Creative destruction. All sorts of talents and skills must have been lost. Mostly communicative. But also things like navigation and probably even the richness of dreams. In the end this strange new code must have replaced at least part of the world with what can be said about it. Reality with opinion. Narrative with commentary.
>
> And sanity with madness, dont forget.
>
> Yes. I wont.
>
> And the arrival of universal war. (175)

"To God every man is a heretic," explains the chapel-keeper in *The Crossing*. "The heretic's first act is to name his brother. So that he may step free from him. Every word we speak is a vanity" (158). There is not space enough to unpack the Cartesian complexities of McCarthy's stance here, although we will return to them in the final chapter; suffice it to say that in the Kekulé essay, he suggests that the unconscious is, foremost, a biological system—"a machine for operating an animal"—and that it can potentially access realities and ideas without attaching language to them. As sculptor Robert Smithson once observed of consciousness, "You don't need existence to exist."[54] To which might be added, a thing or property does not require language to exist—but one of the questions McCarthy raises in the Kekulé essay is, where does the language come from if not the unconscious, which might be exquisitely evolved to its own inscrutable purposes?

Although the Kekulé essay has elicited a good bit of interest and criticism (which McCarthy addressed in a follow-up essay), he had essentially already written it piecemeal in his fiction and certain unpublished writings. For example, in *Cities of the Plain* (1998), in Billy Parham's conversation with the Mexican man beneath the bridge:

It is not the case that there are small men in your head holding a conversation. There is no sound. So what language is that? In any case this was a deep dream for the dreamer and in such dreams there is a language that is older than the spoken word at all. The idiom is another specie and with it there can be no lie or no dissemblance of the truth (281).[55]

This agrees quite exactly with McCarthy's conclusions in the Kekulé essay; what are dreams, if not the unconscious dispensing with language to get our attention about what matters?[56]

The interplay between the unconscious and language in some respect might be said to resemble that of what sculptor Robert Smithson liked to call a dialectical situation. He once explained his *Spiral Jetty* sculpture, fashioned of black rocks and positioned on a remote edge of the Great Salt Lake, this way: "What you are really confronted with in a nonsite is the absence of the site. . . . In a sense the nonsite is the center of the system, and the site itself is the fringe or the edge." *New York Times* writer Heidi Julavits adds, "Sites and nonsite, in other words, involve the equal interplay of consciousness and matter." This might be enlarged to say that the unconscious is implicated in the interplay of site and nonsite, too. Driving on an unfinished stretch of the Jersey Turnpike at night with his friend Tony Smith, Smithson recalled Smith in that moment "describing the state of his mind in the 'primary process' of making contact with matter." "In the same essay," writes Julavits, "[Smithson] noted that Freud referred to this commingling experience as 'oceanic.'"[57] McCarthy's fascination, for example, with whales as an expression of a deeply evolved consciousness in *Whales and Men* explores the possibility that they channel an ancient, "oceanic" unconscious.

McCarthy was intrigued by the possibility that whalesong might express something essential about the deep, prelingual past. In "The Kekulé Problem" he posits that language is non-evolutionary:

> The difference between the history of a virus and that of language is that the virus has arrived by way of Darwinian selection and language has not. The virus comes nicely machined. Offer it up. Turn it slightly. Push it in. Click. Nice fit. But the scrap heap will be found to contain any number of viruses that did not fit.

Darwinian evolution, after all, relies on discards and failed experiments. Poet Katherine Larson, contemplating the sea as the starting place for genetic evolution, sees "Not perfection . . . but originality."[58] Originality is the right term of art because it extends both to the origins of things and the relentless creativity of nature, including its penchant for creative mutation and creative reuse (consider the oft-cited rule, that, with the exception of manatees and certain sloths, all mammals, from giraffes to bats, have seven cervical vertebrae).

Nature can seem heartless in its evolutionary experiments, which result in such phenomena as spontaneous abortions, teratomas, and so-called parasitic twins competing in the womb for a better purchase on life from the moment the zygote splits. The resulting suffering is one manifestation of the theodicy of natural evil and an enduring concern for McCarthy. Suttree suffers from nightmares in which he sees his dead twin, "a sinister abscission," borne on a bier (80). The autopsy of Lester Ballard, the necrophiliac killer of *Child of God*, is a reminder (to turn the tables on Wordsworth) that we dissect to murder, seeking pathologies of the body to explain the sources of evil, which might originate in something as innocuous seeming as a sinister cell. Discovering a tumor in Ballard's brain might give some comfort to those who would explain the mystery of evil in the man. But perhaps it began with the leftward-yawing zygote.

This much is certain about the *Suttree* passage: it is a study in chirality that evokes ancient notions of gauche/sinister/not-right, the word "sinister" appearing a half-dozen times in the text. A cell with left-handed chirality could indeed go own way, and, depending on where it was situated, might go against the grain. Cornelius Suttree is haunted by the possibility that he might be the "evil" twin, in some measure responsible for his stillborn and forever-lost sibling. In the early pages of the text described as

> The ordinary of the second son. Mirror image. Gauche carbon. He lies in Woodlawn, whatever be left of the child with whom you shared your mother's belly. He neither spoke nor saw nor does he now. Perhaps his skull held seawater. Born dead and witless both or a teratoma grisly in form. No, for we were like to the last hair. I followed him into the world, me. A breech birth. (14)

A high percentage of twin births result in head-first presentation of the first, breech of the second. Suttree's stream of thought affirms that his lost brother was his mirror-image twin. Recall the earlier conjecture that a pathological mechanism might be responsible for "both the process of twinning itself and the destabilization of the LR axis." It might be said that genetic coding errors often arise with twins, and some would say that twins are themselves a kind of coding error. But which one was Suttree? The passage is ambiguous. He was the second to be born; does that make his stillborn brother the ordinary, or the gauche (left-handed) human enantiomer, if you like? This much is certain: for much of the novel Suttree feels incomplete and guilt-riddled because of his lost twin, a yin in search of his yang. Upon visiting Immaculate Conception Church, he thinks, "The virtues of a stainless birth were not lost on him, no not on him" (253).

From the very beginning, then, Suttree sees himself in relation to his stillborn twin. The language of the novel is shot through with chirality, in fact, beginning with his recollection of drinking from a mountain spring: "Under the watercress stones in the clear flowage cluttered with periwinkles. . . . A rimpled child's face watching back, a water isomer agoggle in the rings" (13). One does not customarily apply the term "isomer," which refers to molecules with the same chemical formula but a different LR arrangement of atoms, to a human being. But this is a novel about human enantiomers, and enantiomers, after all, are just one type of isomer. Since Suttree sees his mirror-image face, he is perhaps seeing the likeness of his lost brother—a ghost, in other words.

The term "isomer" resurfaces at one other point in the novel, when Suttree encounters two possum hunters who are identical twins, "alike to the crooks in their stained brown teeth" (358). They share a story of chirality and falling simultaneously "out of black walnut trees at the identical same minute eight mile apart. I broke my right arm and Fernon his left'n and he's lefthanded and me right" (360). Suttree will be reminded of his lost twin when the brothers conduct a mind-reading experiment, after which "Suttree stood up. The hunter spun about and faced his unarmed image across the file, his sinister isomer in bone and flesh" (361).

So, once again, we have a dextral and sinistral twin, and the language of chemistry applied to human beings. There are many allusions to astronomy, astrology, and the zodiac in *Suttree*—indeed, McCarthy's notes suggest that they were part of the plan for the novel—and since Vernon and Fernon are hunters, we might ask if they are an unlikely Castor and Pollux. In other words, with their appearance, Gemini strides into the novel.

Why so many twins? Recognizing that doubling, in the literary sense of "splitting" a character into the familiar and the strange, is a stock element of gothic fiction, it might be said that there is something more happening in McCarthy's work with the reconciliation of Suttree and Othersuttree. The quest for wholeness and the self-integration of opposing forces is the

TABLE 2.2 *Symmetric Phraseological Appropriation from Yeats*

Possible source text in Yeats's "Crazy Jane Talks to the Bishop"	McCarthy's reworking of the material in his journalia/notes	Muted allusion in final text of *Suttree*
"A woman can be proud and stiff When on love intent; But Love has pitched his mansion in The place of excrement; For nothing can be sole or whole That has not been rent."	As I grow old will not my spirit mend by rending? And my eyes gather a visionary's light all dusty with loveliness like the sun by morning? Until my heart is an ash charred by charity (torn page)[a]	"Some doublegoer, some othersuttree eluded him in these woods and he feared that should that figure fail to rise and steal away and were he therefore to come to himself in this obscure wood he'd be neither mended nor made whole but rather set mindless to dodder drooling with his ghosty clone from sun to sun across a hostile hemisphere forever" (287).

[a] Cormac McCarthy Papers, Southwestern Writers Collection, The Wittliff Collections, Texas State University-San Marcos, Box 19, Folder 13.

human story writ large, not to say the major religions, and a significant part of the mythic journey per Joseph Campbell, the coming back to the self. Cornelius Suttree comes to wholeness precisely by being split. Consider this textbook example of the sort of "phraseological appropriation," as Michael Crews terms it:

Suttree has done his share of slumming in Knoxville's places of excrement, to be sure. The "ghosty clone," the doppelganger "othersuttree" is Suttree's chiral twin, and the passage shows a strange dependency between the two.

Yet the chief epiphany that Suttree later shares after being sick unto the point of death is this: "I learned that there is one Suttree and one Suttree only" (461). He declares it to an uncomprehending priest from his hospital bed in an overtly heretical cooption of catechistic monotheism and its "one God and only one." It is a radically individualist reclamation of the divine. Yet there are multiple ways to interpret the statement: has Suttree incorporated his enantiomeric twins into a single entity, a whole self? Or has he simply come to terms with chirality, and his identity with respect to his lost twin? In other words, has he let go and accepted his asymmetry? Regardless, it seems Suttree's twins must be separated to be conjoined and reconciled.

Suttree's vision of the end also prompts him to say, "Nothing ever stops moving." Chirality suggests that asymmetry and movement, as much as balance and stasis, are the chief facts of the human condition.

The "Introduction to provocative questions in left-right asymmetry" submits,

> Like symmetry, asymmetry is also pervasive in our aesthetics, where it lends a provocative contrast, representing the dynamic, the unexpected, the emerging and the innovative. Together, symmetry and asymmetry comprise the proverbial Yin and Yang, the black and white, the metaphorical good and sinister. Both are essential for completeness, but too much in either direction disrupts a critical balance—symmetry unchecked by asymmetry transmutes order, harmony and beauty into static, sterile and monotonous. Asymmetry unchecked by symmetry becomes aberrant, unrestrained and chaotic.[59]

Biology itself "holds no exception to this duet." This thinking aligns with Leonard Bernstein's Unanswered Question lectures at Harvard, in which Bernstein points first to our hard-wired duality—"We are symmetrically constituted, dualistically constituted, in the systole and diastole of our heartbeats, the left-rightness of our walking, the in-and-outness of our breathing, in our maleness and femaleness. and love of symmetry"—before entertaining the audience with a Mozart piece he has carefully recomposed for mirror symmetry, with very dull results as it lags in the return to its principal theme. "But what a drag it's been to get there," he admits,

> so much academicism, so many unnecessary schoolboy repeats, such a lack of deletion. This is "prose," or else music by a bad composer. It's stalling for time, the way people do while they're thinking of an answer to a question. "Uh, how old is Mildred?" "Oh, Mildred—must be—uh— Mildred is 69."
>
> Of course my musical repeats were made in the name of symmetry. But symmetry is not necessarily balance: that's a precept we all learned long ago, and it's worth saying again. What Mozart has done—as any great master does—is to make the leap from prosy symmetry into poetic balance, that is, into art.[60]

McCarthy, notes Michael Crews, "was drawn to images of art as either formally static or dynamic" as he worked on *Blood Meridian* and *Suttree* during the 1970s. Crews traces sources in McCarthy's notes and works on this tension, including Wyndham Lewis's *Time and Western Man* (1927), Oswald Spengler's *Decline of the West* (1922), and Michel Foucault's *Madness and Civilization* (1961).[61] In any case, it might be said that this tipping property contains the seeds of creation and destruction. That which spirals is already headed toward its separate destiny, subject to the laws of entropy and gravity. Life persists on earth because the planet has not yet spiraled toward or away from the sun. In the longer view, in Yeatsian terms, things fall apart, and the center cannot hold in an entropic and expanding universe.[62] Indeed, the expanding universe as we know it is made of spiraling stars, gases, and so on, substances that turn and rotate with their own gravity-bound chirality, in some cases spiraling into the compaction of black holes in a cosmic illustration of what it means to circle the drain. Robert Smithson's famed sculpture *Spiral Jetty* on the shores of the Great Salt Lake is a visual rumination on the fact that "the world is slowly destroying itself. . . . The catastrophe comes suddenly, but slowly" (Julavits).[63]

Reflecting on expansion and contraction—the spiraling gyres of the physical universe— in an unpublished fragment of *Whales and Men*, McCarthy writes, "If this looping process is the way it works then there's no need for teleology." As Smithson himself made clear, the *Spiral Jetty* sculpture is as much about creation as destruction. And McCarthy's vision for the great loop of chiral destruction and creation is made clear in *The Road*: "Perhaps in the world's destruction it would be possible at last to see how it was made. Oceans, mountains. The ponderous counterspectacle of things ceasing to be. The sweeping waste, hydroptic and coldly secular. The silence" (231).

FIGURE 2.5 *"The pin has been pulled from the axis of the universe"* (The Crossing 146). *2004 photograph of sculptor Robert Smithson's (right-handed)* Spiral Jetty *(1970), a "profound monument to catastrophe" by* New York Times *writer Heidi Julavits. Credit Wikimedia Commons, Creative Commons Attribution-Share Alike 4.0 International license.*

Here, again, is Gardner's exploration of worlds and antiworlds—"at the extreme limit of the expansion our world will enter a space-time singularity" where contraction begins, the arrows of time reverse, and "The universe, in brief, will turn into a time-reversed world of antimatter" (294). This speculation is not particularly new—as Gardner points out, Socrates speculated on such cycles of reversal in the *Statesman* dialogue. Indeed, the writer's vision is closely akin to Smithson's *Spiral Jetty*; Smithson once described the postindustrial landscape as "ruins in reverse."

Perhaps we might even speculate that our felt sense of profound mythos, and whatever logos that reason may extract from the world, has its underpinnings in the chiral dance of symmetry and asymmetry. Indeed, everything is moving, and we feel it in our deepest substance. Poetry often intuits what science says in a different idiom, and vice versa. If evolutionary theory is a way of exploring the memory of life in the universe, writ large, then poet Katherine Larson wants to show us "Memory. The invention/of meaning. Our minds with deeps/where only symbols creep"[64]— whether they are expressed in mathematics or some other semantic. Synthetic labyrinths, which are often designed as a form of spiritual exercise, generally demonstrate chirality in their winding; the movement transcends semantic meaning.

We, and the stuff we are made of, wander, spin, gyre, and turn. We may long for wholeness and symmetry, but we cannot deny our asymmetry or the tilt and motion of our cosmos, our atoms. McCarthy's imagination has a turn in it, too, and for attentive readers it opens up the mysteries and essential importance of chirality in both the spiritual and material dimensions of the human journey.

FIGURE 2.6 *Stone at the entrance of the Newgrange Tumulus (c. 3200 BC), Ireland, presenting both dextral and sinistral helices. Photographed by the author, July 2016.*

Notes

1   See Martin Gardner, *The New Ambidextrous Universe* (New York: W.H. Freeman and Company, 1990), chapter 18.

2   Ibid., 307.

3   Much of the material summarized here regarding chirality in human development is derived from Michael Levin's "Left-Right Asymmetry in Embryonic Development: A Comprehensive Review" (2004), an article that provides ample context for the science behind chirality, twinning, and what I'll term, conveniently but pejoratively, "malformation"—all matters of some interest to McCarthy that are explored, principally, in *Suttree*, as well as other of his works.

4   Lewis I. Held, Jr, *Animal Anomalies: What Abnormal Anatomies Reveal About Normal Development* (Cambridge: Cambridge University Press, 2021). See especially Chapter 2, "Two-Headed Tadpoles."

5   Lewis I. Held, Jr, and Stanley K. Sessions, "Reflections on Bateson's Rule: Solving an Old Riddle About Why Extra Legs are Mirror-Symmetric," *Journal of Experimental Zoology (Molecular and Developmental Evolution)* 332 (2019): 219–37. doi: 10.1002/jez.b.22910.

6   The scientific debate around asymmetry is alive and well. Cf., Amar Klar, Michael Levin, and Ann Ramsdell, eds., "Provocative Questions in Left–Right Asymmetry," *The Philosophical Transactions of the Royal Society B* 371, no. 1710 (December 19, 2016).

7   Ethan Siegel, "The Universe is not Symmetric," *Big Think*, January 25, 2022, https://bigthink.com/starts-with-a-bang/universe-symmetric/.

8   Anna Kamieńska, *Astonishments: Selected Poems of Anna Kamieńska*, ed. and trans. Grażyna Drabik and David Curzon (Brewster, MA: Paraclete Press, 2011), 79.

9   Louis Pasteur was another pioneer in the concept of chirality, having discovered that differently polarized suspensions of crystals rotate life differently—clockwise and counterclockwise, according to optical rotation. He correctly inferred that something was going on at the molecular level.

10   James P. Riehl, *Mirror-Image Asymmetry: An Introduction to the Origin and Consequences of Chirality* (Hoboken, NJ: John Wiley & Sons, 2011), 59.

11   Ibid., 66.

12   Jason Wallach, "A Comprehensive Guide to Cooking Meth on 'Breaking Bad,'" *Vice.com*, August 11, 2013, https://www.vice.com/en/article/exmg5n/a-comprehensive-guide-to-cooking-meth-on-breaking-bad.

13   Cf. Riehl, *Mirror-Image Asymmetry*, Chapter 3.

14   Andrew Rutherford, "DNA's Twist to the Right is Not to Be Meddled With, So Let's Lose the Lefties," *The Guardian*, April 30, 2013, https://www.theguardian.com/science/blog/2013/apr/30/dna-twist-to-right. See also Katie McCormick, "Asymmetry Detected in the Distribution of Galaxies," *Quanta*, December 5, 2022, https://www.quantamagazine.org/asymmetry-detected-in-the-distribution-of-galaxies-20221205/.

15 Nick Romeo, "Cormac McCarthy on the Santa Fe Institute's Brainy Halls," *Newsweek*, February 12, 2012), https://www.newsweek.com/cormac-mccarthy -santa-fe-institutes-brainy-halls-65701.

16 Michael Levin, "Left-Right Asymmetry in Embryonic Development: A Comprehensive Review," *Mechanisms of Development* 122 (2005): 3.

17 Ibid.

18 Will Edwards, Angela T. Moles, and Peter Franks, "The Global Trend in Plant Twining Direction," *Global Ecology and Biogeography* 16 (2007): 795.

19 Riehl, *Mirror-Image Asymmetry*, 189.

20 Levin, "Left-Right Asymmetry," 3.

21 It remains true, however, that brain asymmetry remains the most compelling explanation for our handedness, since language and fine-motor control are located in the left hemisphere of 97 percent of human brains—and the left hemisphere controls the right hand. On the other hand, fMRI brain scanning is rapidly changing and complicating our understanding of where certain brain functions typically take place.

22 Sarah Huber, email to the author, February 8, 2012.

23 Levin, "Left-Right Asymmetry," 3–4.

24 Mikiko Inaki, Jingyang Lie, and Kenji Matsuno, "Cell Chirality: Its Origin and Roles in Left-Right Asymmetric Development," *The Philosophical Transactions of the Royal Society B* 371, no. 1710 (December 19, 2016): 8.

25 Pam Belluck and Tania Franca, "Clues to Zika Damage Might Lie in Cases of Twins," *New York Times*, May 1, 2017, https://nyti.ms/2po6RqX.

26 Levin, "Left-Right Asymmetry," 15.

27 Ibid.

28 Riehl, *Mirror-Image Asymmetry*, 190. The percentage of left-handers in the mentally developmentally delayed population is 20% (vs. 10% in the population at large), and among the severely delayed, 28%. Ibid, 189.

29 Annie Dillard, *Teaching a Stone to Talk: Expeditions and Encounters* (New York: HarperCollins, 2009), 19–20.

30 Cormac McCarthy Papers, Southwestern Writers Collection, The Wittliff Collections, Texas State University-San Marcos.

31 Michael Lynn Crews, *Books Are Made of Books* (Austin, TX: University of Texas Press, 2017), 196–8.

32 Ibid.

33 Riehl, *Mirror-Image Asymmetry*, 72.

34 Curiously, this tendency against purely racemic distribution of compounds is itself a feature of living things. And the fact that chirality of compounds changes very gradually after death is what allows for the carbon dating of organic materials.

35 David Tracy, *The Analogical Imagination: Christian Theology and the Culture of Pluralism* (New York: Crossroad, 1981).

36   I am speaking here of ordinary time in human experience, without reference to relativity or the five (or more) arrows of time that physicists describe. Interestingly, the laws of nature are symmetrical with respect to time, with the notable exception of the second—entropy is conserved.

37   Per Merriam-Webster's second register of *logos*: "reason that in ancient Greek philosophy is the controlling principle in the universe."

38   Drafts of *Whales and Men*, Southwestern Writers Collection, The Wittliff Collections, Alkek Library, Texas State University-San Marcos, Cormac McCarthy Papers, box 97, folder 5.

39   John Henry Holland, *Complexity: A Very Short Introduction* (Oxford: Oxford University Press, 2014), 11.

40   Gardner, *The New Ambidextrous Universe*, 291–2.

41   Samuel Baron, Kristie Miller, and Jonathan Tallant, *Out of Time: A Philosophical Study of Timelessness* (Oxford, 2022); Giulia Rubino, Gonzalo Manzano & Časlav Brukner, "Quantum Superposition of Thermodynamic Evolutions with Opposing Time's Arrows," *Communications Physics* 4, no. 251 (November 26, 2021), doi: https://doi.org/10.1038/s42005-021-00759-1.

42   Quoted in *Stories Are Made of Stories*, 272. Michael Crews writes of this passage, "McCarthy's gestalt cosmology derives from multiple sources, but the archives confirm Teilhard de Chardin is one of them." Of course, the singularity of stories has long been a matter of intrigue for literary critics and psychologists. For another entry in this vein, see Christopher Booker's *The Seven Basic Plots: Why We Tell Stories* (London: Continuum, 2004).

43   For his part, Martin Gardner, a masterful logician and investigator of all things chiral, conceded, "I don't think there's any way to prove the existence of God logically." Quoted in Phillip Yam, "Profile: Martin Gardner, the Mathematical Gamester (1914–2010)," *Scientific American*, May 22, 2010, https://www.scientificamerican.com/article/profile-of-martin-gardner/.

44   Garry Wallace, "Meeting McCarthy," *Southern Quarterly* 30, no. 4 (1992): 138.

45   McCarthy uses the term "logos" repeatedly in both his unpublished and published writing. For more on this see Bryan Giemza, "Last Roads Taken: Robert Frost, Cormac McCarthy, and Dying Worlds," in *Undead Souths*, ed. Eric Gary Anderson, Daniel Cross Turner, and Taylor Haygood (Baton Rouge: LSU Press, 2015), 161–72.

46   As articulated in Thomas Hobbes, *Leviathan* (Oxford: Oxford University Press, 2008).

47   See also Dianne C. Luce, "Cormac McCarthy in High School: 1951," *The Cormac McCarthy Journal* 7, no. 1 (2009): 1–6, http://www.jstor.org/stable/42909394.

48   The laws of thermodynamics play a part in this as well. Viewed in the long arc, we would expect systems to move away from complexity and organization. David Layzer's well-known Arrow of Time argument attempts to reconcile the general trend (movement toward chaos) with spurts of increasing order.

49   Charles Darwin, *The Descent of Man and Selection in Relation to Sex* (New York: P.F. Collier, 1901), 178.

50 Cormac McCarthy, "The Kekulé Problem: Where Did Language Come From?," *Nautilus* 47 (April 20, 2017), http://nautil.us/issue/47/consciousness/the-kekul-problem.

51 Crews, *Books Are Made of Books*, 270.

52 Drafts of *Whales and Men*, Southwestern Writers Collection.

53 *Whales and Men* [screenplay], n.d., final draft, printout with no corrections, Cormac McCarthy Papers, box 97, folder 5, 57.

54 Quoted in Julie Julavits, "The Art at the End of the World," *NY Times Magazine*, July 7, 2017, https://www.nytimes.com/2017/07/07/magazine/the-art-at-the-end-of-the-world.html.

55 Compare Samuel Beckett's 1949 postmodern declaration of artistic creed to Georges Duthuit as "The expression that there is nothing to express, nothing with which to express, nothing from which to express, no power to express, no desire to express, together with the obligation to express." *Proust [and] Three Dialogues with Georges Duthuit* (London: Calder and Boyars, 1965), 103. The follow-up response essay reference is "Cormac McCarthy Returns to the Kekulé Problem," *Nautilus*, November 27, 2017, https://nautil.us/cormac-mccarthy-returns-to-the-kekul-problem-236896/.

56 Case in point, from the essay: "Those disturbing dreams which wake us from sleep are purely graphic. No one speaks. These are very old dreams and often troubling. Sometimes a friend can see their meaning where we cannot. The unconscious intends that they be difficult to unravel because it wants us to think about them. To remember them. It doesnt say that you cant ask for help. Parables of course often want to resolve themselves into the pictorial. When you first heard of Plato's cave you set about reconstructing it."

57 Julavits, "The Art at the End of the World."

58 Katherine Larson, "Ghost Nets," *Radial Symmetry* (New Haven, CT: Yale University Press, 2011), 41.

59 Klar, Levin, and Ramsdell, eds. "Provocative Questions in Left–Right Asymmetry," 1.

60 Leonard Bernstein, "Lecture 2: Musical Symmetry," Television Script, *The Unanswered Question* (Six Talks at Harvard), 1973, https://leonardbernstein.com/lectures/television-scripts/norton-lectures/musical-syntax.

61 Crews, *Books Are Made of Books*, 108.

62 As articulated in *A Vision* (1925), W. B. Yeats was fascinated by gyres, which are chiral coils.

63 See Smithson's 1972 essay, "The Spiral Jetty," https://www.diaart.org/media/_file/brochures/spiral-jetty-7-18-for-website-3.pdf. In Jeffrey Kosky's interpretation, "[Smithson] says he was seeking 'red'—described by G.K. Chesterton in the essay's epigram as 'the most joyful and dreadful thing in the physical universe . . . the place where the walls of this world of ours wear the thinnest and something beyond burns through.'" Jeffrey L. Kosky, "Learning to Live on the *Spiral Jetty*," *Image* 84 (October 28, 2017), https://imagejournal.org/article/learning-to-live-on-the-spiral-jetty-2/.

64 Katherine Larson, "Ghost Nets," 41.

3

Technology

Blowing Up Knoxville–
How Domestic Terrorism
and Actual Misadventures
with Dynamite Shaped
McCarthy's World

On January 29, 1958, a junior high school student, Charles Pruitte (15), accompanied Glenn Gulley (17), a high school dropout, on a country ramble. Armed with their .22 rifles, looking for "rabbits and cans to shoot," they left their housing project in Lebanon, TN (population then approximately 10,000) and eventually found their way to a grocery owned by H. A. Lasater, who saw them come in for a cold drink.

"He drank it right down and went out," Lasater later said of Gulley, with Cormackian concision.[1]

Gulley did not carry his rifle into the store, but he must have had it, as well as a flash of inspiration, when a short while later the lads came across a trailer set off by a mile from the county courthouse, as Murfreesboro's *Daily News Journal* noted with supercilious scare quotes, "for safe-keeping." Marion Construction Company was working on a nearby road construction project and had left a trailer loaded with (company officials claimed) 1,500 pounds of dynamite in a portable magazine in the middle of a field, "away from everything," according to Malcom S. Poteat, the owner of the company, "as a safety measure." He added that it was clearly marked with the word "dynamite" in chalk lettering.

What do teenage boys with rifles do with such a thing?

They shoot at it, obviously. In this case, four or five times, according to later reports.

At 2:15 PM the trailer detonated in an explosion heard as far away as Lafayette, some twenty miles distant.[2] Glen Gulley was killed instantly, his body found 75 feet from the trailer. Some of the twisted wreckage of the trailer landed 250 yards away. The concussion knocked town residents off their feet. Mrs. Otis Johnson was "carrying an iron[ing] board into her kitchen, and reported that, 'It knocked the ironing board out of my hand and threw it across the room. It knocked me onto the floor. I didn't know what happened.'" Twenty-five state troopers descended on the area. Reports of property losses, including those from around twenty town center

FIGURE 3.1 *Front page of* The Nashville Tennessean, *morning edition, January 30, 1958, headlining the explosion in Lebanon, TN. Note that the* Tennessean *identifies itself "At the Crossroads of Natural Gas and Cheap TVA Power." Cormac was twenty-five at the time of the incident. The day's edition also carries news of the capture of an infamous western "mad dog": teenaged killer Charles Starkweather.*

businesses, were filed from locations as far as two miles away, as windows throughout town, including some in the courthouse, were shattered, and several families were displaced from their homes.[3] By the next day it had been discovered that many area homes had been shaken from their foundations with insurance losses tallied at around $2,000,000 in today's dollars.[4]

A smoldering crater the size of a house foundation remained to mark the spot of the escapade.

Missing an eye, in tatters, half naked, with a broken arm and a puncture wound extending completely through his abdomen, young Charles Pruitte ran 200 yards to collapse in the arms of a passerby. Pruitte's later comment in hospital to a reporter would satisfy Gene Harrogate, the scrappy miscreant of McCarthy's *Suttree*, after his nearly fatal exploits with dynamite in subterranean Knoxville: "I fired the last shot, and then I don't know what happened."[5]

Subsequently the town fire marshal identified many safety infractions that contradicted Poteat's account: the trailer had not been designated as a storage magazine, was in fact located 200 feet inside city limits (and 534 feet from the nearest house), and was found to have contained as much as "5,000 pounds of dynamite at one time"—not the 1,500 originally claimed. Additionally, the marsha l discovered two cases of dynamite belonging to the Marion Company "stacked in front of a [Lebanon, TN] grocery store where customers were going in and out." The boys, he said, had not known that it was a dynamite trailer, but fired at the trailer "just to be shooting at something."[6]

Dynamite was typically sold in cases weighing approximately fifty pounds, the maximum personal allowance for civilians. Poteat explained that in the course of ordinary construction his crews would use about five cases a day—more for a "heavy rock job." A "mere" fifty pounds of dynamite for home use (22.7 kg) would release about 70 percent of the electric energy equivalent of a typical American household's daily usage—in a flash. In all likelihood, the boys released a hundred times that amount of energy in an instant.

In an era before the advent of the news chyron, print media experts eagerly dissected the Lebanon incident from every angle. Grisly photographs of the scene ran in multiple newspapers, including a gratuitous front-page picture in the *Tennessean* showing Gulley's crumpled body at the scene, his side blown open and his skull denuded of flesh. The explosion's object lesson raised singed eyebrows around the state, and the day after the incident the *State* Journal, after consulting legal authorities, concluded that "Tennessee apparently has no dynamite storage laws less than 100 years of age."[7] Per State Code Section 39-1404, adopted in 1848 for the storage of gunpowder and explosives, no one could possess more than fifty pounds of explosives within town limits, with violations punishable by a fine "not less than $100" with a presumption of liability for damages. The Lebanon case fell into something of a gray area: another set of dynamite laws recently enacted by the State General Assembly pertained narrowly to the use of "explosives for malicious purposes." The legislative action was no accident. In the run-up to *Brown v Board* it was becoming increasingly clear that dynamite was the instrument of choice for domestic terrorists, especially in the latterday vigilante enforcement of Jim Crow in the southern United States.

This was consistent with a long history in the southern Appalachian Mountains of using dynamite for settling scores, as happened during the terrifying night hours of October 5, 1958, when Alfred Nobel's technology, which had not yet celebrated its centennial, was used by unapprehended cowards to resist school integration by rendering Clinton High School into a smoking ruin. Displaced students would be relocated to the "neutral" federal facilities of nearby Oak Ridge; Cormac was attending college at the nearby University of Tennessee. Newspapers of the era show dynamite turning up throughout Tennessee and the region constantly, under murky circumstances, and often without a chain of provenance. Throughout the South, dynamite, long treated as a common farm implement for land clearing and other routine chores, enjoyed a sort of plausible innocence. Laxly regulated and widely available in hardware stores across Tennessee, it was not until the Organized Crime Control Act of 1970 finally placed the regulation of explosives regulations under the purview of the Bureau of Alcohol, Tobacco and Firearms (ATF) that the sun set regionally on the freewheeling personal possession of dynamite.

Steven Johnson, a popular science writer known for probing the history of ideas and technologies, offers no entry for dynamite in *How We Got to Now: Six Innovations That Made the Modern World* (2015). He devotes its pages instead to such taken-for-granted innovations as glass, air-conditioning, clocks, and water purification, and especially the long train of their unintended consequences.[8] Since most innovations are swiftly applied to warfare (e.g., the invention of aircraft immediately followed by their use to deliver explosives and conduct surveillance) if not created by warfare, perhaps it is so obvious as to merit little discussion, but it bears underscoring nonetheless: petrochemical agricultural systems, sped along by the Second World War, have had the net effect of accelerating the disruption of global carbon, nitrogen, water, and climate systems. In the short run, they made it possible to feed a global human population explosion. These liabilities and assets have been well examined in popular works such as Michael Pollan's *The Omnivore's Dilemma*.[9]

Less attention has been devoted to what Patrick Benjamin Swart calls "the golden age of chemical high explosives in American agriculture." In the history of things and transformative technology, no small part of dynamite's history must be writ in a sort of military-industrial agriculture that created mutually reinforcing relationships between mechanization and concentrated energy. In his useful thesis, *Blasting the Farm: Chemical High Explosives and the Rise of Industrial Agriculture, 1867-1930*, Swart summarizes,

Invented in 1867 by Swedish chemist Alfred Nobel, dynamite, one of the earliest and most common chemical high explosives, marked a distinct shift in the history of explosive power. Chemical high explosives' utility resulted from the fact that explosives straddle the line between energy and mechanization. Much like coal and oil, high explosives' value emanated from the potential energy stored in their chemical bonds. Unlike these other forms of energy, explosives did not require an external apparatus, such as steam or internal combustion engines, to harness their power. Instead, the mechanical forces of dynamite transferred directly, by way of rapidly expanding gasses, to whatever solid matter it encountered. In the second half of the nineteenth century, manufacturers of explosives primarily marketed dynamite, nitroglycerine and other high explosives to the mining and construction industries. But in this same period, farmers gradually began to harness its "concentrated power." By 1920, farmers regularly used chemical high explosives in agriculture.[10]

Which is to say that fantasies of ecosystem control were conveniently reduced to scale and the possibility of quite literally removing whatever stood in the way. In that sense, dynamite is to old blasting techniques what the jacketed bullet was to powder and ball. Joseph Husband, whose brief and eventful life took him to the advertising world of Chicago as well as the navy during the First World War, began his career by journeying immediately from his Harvard graduation to midwestern coalmines, where he served as a correspondent for the *Atlantic Monthly*. He explains,

> In the first days of coal-mining . . . a man did all the work. With his hand-drill he bored into the face of the coal at the head of his room, or entry, and from his keg of powder he made long cartridges and inserted them into his drill-holes. Then, when the coal was down, and he had broken it with a pick, he loaded it with his shovel in a car.[11]

Of course, this process was more fully mechanized and electrified even by Husband's time. Yet his 1910 account of modern techniques and equipment would be recognizable in its essentials to miners today. At Albuquerque's Nuclear Museum, one learns about how nuclear warheads were miniaturized to the dimensions of rocket-propelled grenades and suitcases, examples of the level of control in so-called battlefield nukes (tactical or nonstrategic nuclear weapons). SFI founders who were involved in the Manhattan Project had brought the nuclear age to the mountains of Tennessee and to the world before eventually scaling it down to the palm of the hand.

McCarthy asks, quite pointedly in *Suttree*, whether Knoxville's penchant for building atop the necropolis of its own ruins can ever generate meaning in a culture fueled by relentless fire (including dynamite) and ephemeral visions of progress in a zero-sum game of invention and destruction. As climate communicator Bill McKibben pointed out recently in *The New Yorker*, carbon neutrality based on any combustive process simply moves sums around the carbon spreadsheet, and we now understand that burning biomass, for example, is hardly carbon-neutral. One way to conceive of this is by considering that fire stills rules the world (appallingly so in our time of climate change), its energy splitting the globe, and even (indirectly) the atom and the epochs, a primary theme of McCarthy's Border Trilogy explored in the next chapter. Referring to the evolutionary burst that fire conferred to the human brain, McKibben concludes, "Since the large brain originally underwritten by those fire-cooked meals has figured out how to take advantage of that distant force [of the sun] . . . we can, and must, bring the combustion age to a swift end."[12] Energy in the form of wood, charcoal, coal, uranium—all require the application of additional energy to be concentrated. In a remarkable way, the Tennessee Valley Authority's (TVA) efforts toward hydroelectric power ran against the petrochemical dependencies that erected its dams. Still, McKibben's bottom line is inescapable: most forms of renewable energy draw sustenance from the sun, including biomass, wind energy, and most hydropower (with the exceptions of tidal energy and geothermal sources).

Critics have debated whether it is nostalgic or anti-nostalgic, but certainly *The Orchard Keeper* (1965) harkens to the shoulders of those industrial giants and frontiersmen who carried modernity forward. For example, Uncle Ather and Warn's grandfather cut railroad "sleepers for the K S & E," aka the shortline known as Knoxville, Sevierville and Eastern, but more commonly referred to locally as the Knoxville Slow and Easy. The anachronistic rail line, still operational during McCarthy's youth, was nationally renowned among trainspotters eager to spy its wooden cars, pulled by a relic steam engine, crawling a once-daily round trip along deteriorating tracks. While Charles McCarthy, Senior, touted the progress of the TVA, a living shrine to the golden age of railroad industry was plugging along in the background, and in fact the slow-but-steady energies of its little engine were turned to account for the building of Douglas Dam across the French Broad River in Sevier County, Tennessee, to meet the needs of the American war effort. Like a sterile mule (a form of biomass) hauling timber from a virgin southern forest (biomass), the S & E, which finally ceased operation in 1961, harnessed ancient energy as an unlikely servant to its own arid enterprise and obsolescence.

How is it possible to keep entropy and collapse at bay, given the promiscuous use of nonrenewable resources? SFI's Geoffrey West has an answer, but it requires a cartoonish scramble on an ever-accelerating technological treadmill. "We have sustained open-ended growth and avoided collapse," he writes, "by invoking continuous cycles of paradigm-shifting innovations such as those associated on the big scale of human history with discoveries of iron, steam, coal, computation, and, most recently, digital information technology. Indeed, the litany of such discoveries both large and small is testament to the extraordinary ingenuity of the collective human mind."

Unfortunately, however, there is another serious catch. Theory dictates that such discoveries must occur at an increasingly accelerating pace; the time between successive innovations must systematically and inextricably get shorter and shorter. For instance, the time between the "Computer Age" and the "Information and Digital Age" was perhaps twenty years, in contrast to the thousands of years between the Stone, Bronze, and Iron ages. If we therefore insist on continuous open-ended growth, not only does the pace of life inevitably quicken, but we must innovate at a faster and faster rate. We are all too familiar with its short-term manifestation in the increasingly faster pace at which new gadgets and models appear. It's as if we are on a succession of accelerating treadmills and have to jump from one to another at an ever-increasing rate. This is clearly not sustainable, potentially leading to the collapse of the entire urbanized socioeconomic fabric. Innovation and wealth creation that fuel social systems, if left unchecked, potentially sow the seeds of their inevitable collapse. Can this be avoided or are we locked into a fascinating experiment in natural selection that is doomed to fail? (31)

* * *

In *Suttree* (1979), juvenile delinquent Eugene Harrogate takes Cornelius Suttree to inspect a sodden wooden shoring wall that when prized away reveals a bank building foundation: "After a while he raised his head. Dynamite, he said" (263). When Suttree later asks Harrogate about his obsession—"You breached the bank vaults yet?"—Harrogate waves him inside the little concrete bunker where he squats.

> Looky here, said the city mouse.
> What is it?
> Suttree was kneeling. He reached into the dark and felt a wooden box where cold waxed shapes like candles lay. He lifted one out and turned it to the light.
> Gene, you're crazy.
> That's the real shit there. Buddy boy that'll get it when Bruton Snuff wont.
> You can't blow it. You dont have a detonator.
> I can blow it with a shotgun shell.
> I doubt it.
> You keep your old ear to the ground.
> Gene, you'll blow yourself up with this shit.

> I thought you said I couldnt blow it?
> Suttree shook his head sadly. (263)

This goes about as well as readers would expect:

> When Harrogate pulled the string on his homemade detonator he had one finger in his ear. The explosion blew him twenty feet up the tunnel and slammed him against a wall where he sat in the darkness with chunks of stone clattering everywhere about him and his eyes enormous against the unbelievable noise in which he found himself. Then he was sucked back down the tunnel in a howling rush of air, his clothes scrubbing away and peeling of hide until he found himself lying on his face in the passage with a shrieking in his ear. Before he could rise it returned and snatched him up again and scuttled him back along the floor in a cloud of dust and ash and debris and left him bleeding and halfnaked and choked and groping for something to hold to. (269)

Although teenaged Eugene Harrogate is below ground at the time of the incident—hence his being snatched up by the back blast in the cave chamber—the description gives a sense of what Charles Pruitte might have experienced after he pulled the trigger, had those memories not been knocked mercifully from his head. As soon as Suttree reads the day's newspaper headline ("Earthquake?"), he infers what has happened, and after four days of subterranean sorties, he finally locates Harrogate, a process that reverses the roles of Harrogate's poignant earlier search for Suttree after his time served. The bowels of the underworld have been shaken. Among the unintended consequences of Harrogate's exploit are a team of spirits unleashed and the breaching of a sewer line, engulfing him "in a slowly moving wall of sewage, a lava neap of liquid shit and soapcurd and toiletpaper," making his ensuing captivity particularly miserable (270).

When Suttree rescues the dehydrated adolescent four days later, in the subterranean cell "their faces were blackened like miners or minstrels and the city mouse wore only shreds of clothing and he was covered with dried sewage. True news of man here below" (276). The phrasing expressly identifies Harrogate as a nave of misery stuck in a real-life minstrel show. Literary critics have strained to find a sufficiently overarching formula to describe all the violence in McCarthy's work—whether it be sacramental, purifying, or creative—but in this instance, at least, he connects most obviously to the southwestern humor tradition in which the hapless suffer violent injury not merely from their witlessness but very often from American arms wielded to enforce racial and social hierarchies. Class helped to define the parts, often with an educated interlocutor (e.g., Cornelius Suttree) eliciting tales of misfortune from his social inferior (Harrogate), a script that came to define the minstrel show as well. While the southwestern humor tradition is traceable to antebellum southern literature, vestiges of it are familiar in the pratfalls of American popular cartoons of the twentieth century (Elmer Fudd, for example), in which grievous bodily harm, torture, dismemberment, immolation, and lynching are played for laughs.

The poor nutritional fare reserved for the American lumpen applies to their thin diet of humor and culture, as well. During his flush days after killing a stolen shoat Harrogate would lie "these last warm days in a den in the honeysuckle and read comicbooks he'd stolen, risible picturetales of walking green cadavers and drooling ghouls" (144). After he tries to seduce the younger of two "nubile young black girls" to his comic book bower, the sisters "carry off his whole supply." It is a retelling of the ancient ribald tale of the farmer's daughters, yet in this instance the African American father recognizes Harrogate's pathetic circumstances through a common, pan-racial culture of the desperate and discarded, demonstrating remarkable compassion and restraint in allowing Harrogate to work off his debt for the stolen hog. Harrogate's efforts to blast his way into the bank vault find some substantiation in history, too. In 1900 burglars in mountainous Rogersville, Tennessee, dynamited two safes and made off with around $1,350 in cash, bloodhounds in pursuit.[13] In 1904, burglars in Roanoke, North Carolina, spirited away $3,000 after dynamiting a bank safe.[14] True to form, Harrogate is using yegg techniques that are a throwback to fifty years earlier and the stuff of an adolescent's comic book imagination.

Setting aside any easy presentist moral outrage, let's grant that tales of dynamite wildcatting, ghosts, and cruel racist pranks are mainstays of the Southwest Humor tradition. This chapter frames the matter in the negative, insofar as it points to the destructive animation of racist ideology, resistance to state control, and even the sinister tint of popular humor. Let it be said that joy itself is an act of resistance (consider Suttree's moments of ecstasy in isolation and as a true social dropout), and resistance need not be violent—even if violence is the stock of Cormac's fiction. One of the narrative tricks *Suttree* plays on readers is its Lethe-stream of humor that can make us forget how badly Knoxville uses its people and, simultaneously, their resistance to being used. Fittingly, outsized southwestern tales offer a notably mean-spirited sort of humor tradition as a demonstration of force for the weapons of the supposedly weak, and they point also to the taproot of the not-so-mysterious American "mystery of violence." A proclivity for life-threatening escalations in a land economically defined by enslaved or indentured labor, and that required genocidal displacement for settlement, is not surprising. Perhaps less consideration is given to how American "humor" attempted to render ferocious traditions more "innocent" and acceptable—to debarb and transmit them, in fact—or, in other words, to carry out the so-called cultural work of lowering collective revulsion toward violence and especially race- and class-related conflict. Such was the project of southwestern humor, which bled into minstrelsy, American popular theater, and by the time of young Charles "Cormac" McCarthy, Jr., popular radio and television entertainments in which dynamite, not accidentally, was a standard prop.

If dynamite serves as a comic instrument of a particular kind in *Suttree*, it is equally an instrument of loss or at least incomplete deliverance, the sort of creative destruction that Cormac McCarthy favors elsewhere in his opus. In *The Orchard Keeper*, the youths John Wesley, Warn, and Johnny Romines trade tales in a cave, prompting Romines to describe how he wired a "dynamite cap stolen from the quarry shack" to a model train transformer. After luring birds to ground with birdseed following a heavy snowfall, the adolescent blows up the flock:

> Goddamn but it come a awful blast, said Warn. I eased the switch on over and then BALOOM! They's a big hoop of snow jumped up in the yard like when you thow a flat rock in the pond and birds goin ever which mostly straight up. I remember we run out and you could see pieces of em strung all out in the yard and hanin off the trees. And feathers. God, I never seen the like of feathers. They was stit fallin next mornin. (149)

Such exploits were commonplace around Knoxville, of course. At the 2015 Cormac McCarthy Conference Society I spoke with two Knoxville natives, roughly Cormac's contemporaries, who described routinely bringing shotguns, blasting caps, and dynamite caps to show-and-tell during their schooldays (items that the teacher kept in a cabinet for return at the end of the day). Such was the price of industry that Knoxville newspapers reported in 1952 the fate of "Chucky" Mullins, who sustained "eye and leg injuries suffered when he exploded a dynamite cap" he found on the way from school.[15] A suit for $85,000 in damages resulted after young James Donelly Phillips picked up a blasting cap at a Knoxville apartment construction site "and kept it for 10 days before connecting it to a battery," critically injuring him.[16] Eight-year-old Ray Ball of nearby Sevierville lost three fingers to a blasting cap the next year while out of school. "He'd been sick," his mother said, "and you know how it is; you let them do what they want to placate them. He had been climbing up in a cabinet and must have found it up there."[17]

Not only children were wounded in such incidents. In 1982, Sammy Carroll, a retired coal miner with an inveterate wariness of dynamite, ironically suffered severe injuries after apparently unearthing a cache of old caps in the woods behind his house in Devonia, a defunct mining community near Knoxville (with McCarthyesque stoicism the wounded man, missing part of his hand, proclaimed, "Ain't no one going to come and help me. I'm going to bleed to death").[18] Growing up in the 1980s, I remember a close childhood friend in North Carolina, who, under his older brother's tutelage, made pipe bombs using fertilizer, charcoal, urine, and shotgun shell caps in a fashion not unlike *Blood Meridian's* Glanton gang. Ringed by their enemies, they huddle before the malpais, improvising their gunpowder from charcoal, saltpeter, sulfur, and urine ("and the judge on his knees kneadin the mass with his naked arms and the piss was splashin about and he was cryin out to us to piss, man, piss for your very souls for cant you see the redskins yonder, and laughin the while and workin up this great mass in a foul black dough, a devil's batter by the stink of it" [138]). My friend's father described, in what I suspect is the rural cousin of urban legend, the stealing of chemistry lab sodium for the sake of floating it down a local river in a sort of moveable fireworks show. We were taught in high school chemistry that the results were a simple chemical reaction, yet, as science explainer Ethan Siegel writes,[19] "It's easy to say, 'it's just a chemical reaction,' but this reaction, at an atomic and molecular level, is governed by the rules of quantum physics." As a teenager I played a nighttime game of tag with friends wielding Roman candle fireworks; observing safety first, we sensibly donned workshop goggles. In the early 1990s, my next-door neighbor in Raleigh created a homemade "potato cannon," fashioned of PVC pipe and tamped with a hairspray igniter, that more often than not atomized its airborne charges before they could reach the skies above our cul-de-sac, while a neighborhood full of newly suburbanized Vietnam veterans applauded in admiration.

In short, the American South that I grew up in, and that McCarthy grew up in, was one in which white southerners demonstrated a continuing love affair with explosives, from which one might spin all sorts of theories of suppressed, ritualized violence and excitement, impotence and fear, and robust regional devotion to war campaigns following the Civil War as part of a paramilitary culture that could not, or would not, perceive itself as fully militant. Rendering these volatiles into the stuff of frolics effectively decoupled them from being seen as weapons of persuasion, intimidation, and terror. City manager and eventual mayor, George Dempster wrote for the *Knoxville Journal* a playful 1930s encomium to Knoxvillian Tom McFarland, "sire of the pedagogic McFarland girls of the city school system," who

> always carried a half dozen or more blasting caps in his vest pocket, whether on or off work. . . . The caps could easily have been exploded by a sharp blow or a burning match and the damage would have been equal to a small hand grenade, yet Mr. McFarland was in his deep seventies when he died peacefully at this Fifth avenue Knoxville home.

Dempster explained how the old charges were often fused by crimping between the teeth, "a practice that sometimes brought decapitation," and recalled a tale from across the mountain range in neighboring North Carolina where a construction "camp 'flunkey'" inadvertently swept blasting caps into a pot-bellied stove. His culminating anecdote involves "a popular pastime at the old Peabody school on Morgan Street in Miss

Lillian Renfro's room," throwing "one blasting cap or an .38 cartridge at a time into the stove with the waste paper." The expelled students reprised the trick at another school and eventually met the wrath of "the brilliant Mrs. Powers and received resounding whacks over concrete heads with a long heavy pointer for their pains."[20]

The tone is pure George Washington Harris, the grandfather of southwestern humor, remembered for the violent comic capers of Sut Lovingood, a character Jack Neely describes as "Huck Finn on amphetamines, a manic, perverse child of some backwoods holler where Idiocy and Genius fuse into one."[21] This description might apply equally to the hapless Eugene Harrogate, who abounds in schemes confounding in their combination of untutored internal logic and mad foolishness. As Harrogate cannot find shelter as either a country mouse or a city mouse, he is heir to the worst of both worlds in modernity and is schooled in a world of troubles guiding him naturally to hedonistic conclusions in step with Sut Lovingood's. In life, G. W. Harris cast his lot with white supremacists in the Anglo-American way, profiting from the shipping lanes of Cherokee removal and aligning himself with pro-Confederate Nashvillians as the American Civil War approached. The hillbilly character provided a tough, white skin for the hard-luck insouciance of minstrelsy, much like Joel Chandler Harris's animals and later Irish American tales gave space for comic transference, making it possible to attribute some measure of superiority even to the maleducated Poor White, otherwise indistinguishable from his African American counterpart, per the Dukes of Hazard motto: just a good ole boy, never meaning (much) harm.

Especially during the Appalachian phase of his novels, McCarthy would have every reason to know Harris's work. Harris's long association with Knoxville, where he worked on riverboats and as a prospector, ties directly to *Suttree*, as does a literary lineage spanning from Mark Twain, thence to William Faulkner, and on to Flannery O'Connor—three writers of great significance in McCarthy's earlier works, each exploring, in their way, the horrific legacies of American racial apartheid and lingering fears of class- and race-impurity. The dialogue just before the bird-blasting incident is recalled—in a cave, no less—is at once instantly familiar and revealing for its Huck Finn/Tom Sawyerish unwriting of American racial naivete:

> Which you'd rather be, Boog asked John Wesley, white or Indian?
>
> I don't know, the boy said. White I reckon. They always whipped the Indians.
>
> Boog tipped the ash from his cigarette with his little finger. That's so, he said. That's a point I hadn't studied.
>
> I got Indian in me, Johnny Romines said.
>
> Boog's half nigger, said Warn.
>
> I ain't done it, Boog said.

You said niggers was good as whites.

I never. What I said was *some* niggers is good as *some* white is what
I said.

That what you said?

Yeah.

I had a uncle was a White-Cap, Johnny Romines said. You ought to
hear him on niggers. He claims they're kin to monkeys.

John Wesley didn't say anything. He'd never met any niggers. (137)

In a celebrated trial of the late 1890s, east Tennessee Judge T. A. R. "Tar"
Nelson, "recognized as one of the greatest criminal jurists in Tennessee"
by the account of a contemporaneous newspaper, successfully brought to
justice two murderers, Pleas Wunn and Catlett Tipton, who emerged from
the White Cap stronghold in Knoxville's adjacent Sevier County.[22] Nelson
withstood personal threats and a rigged jury initially stacked with several
additional members of the White Caps.[23] Racial terrorism of nineteenth-
century secret societies, pitting rural whites against their Black counterparts
in contention for class mobility and political power, had many permutations
in the American South beyond the Ku Klux Klan, including the Night Riders
and the Red Shirts. Their variants include racial cleansing pogroms such as
the Wilmington, North Carolina, and Tulsa, Oklahoma, massacres, and, in
McCarthy's time, the Clinton, Tennessee, integration riots and high school
dynamiting, and closer to the present era, the Unite the Right Rally in
Charlottesville, 2017. In fact, McCarthy's father counted Clarence Darrow,
one of the Scopes Trial attorneys, among his colleagues. Young Cormac
grew up in direct listening distance of the reverberations of racially divisive
pseudoscientific and religious debates regionally centered in Knoxville. (Such
tales are echoed in *The Orchard Keeper* when Warn points out an African
American church and homestead abandoned after one of his relatives "run
em ever one off"; his son, who favors his father in violence, "jest got out
of Brushy Mountain" penitentiary.) The adolescent discussion of race in
The Orchard Keeper is followed immediately with a tale of "harmless,"
explosive violence.

As the embodiment of a certain sort of hillbilly hero, Sut is an anti-
authoritarian who specializes in revenge pranks to match his own pratfalls.
The mishaps suffered in his tales include sexualized aggression, scatological
humor, degrading nudity, swarms of bees, and the like, and in a seemingly
unlikely modern survival, the tradition continues unabated in the self-
inflicted harms of Johnny Knoxville's *Jackass* series or the defiant, self-
deflating idiocy of southern machismo in Danny McBride's *East Bound and
Down* HBO series (2009–13). A cruel turn in rustic bawdy humor, of course,
is itself of ancient lineage, whether in Geoffrey Chaucer's *Canterbury Tales*
or Giovanni Boccaccio's *The Decameron*. Folk tales from the British Isles

took on an American Irish accent in Appalachia through the Jack Tales, including recent entries in the tradition such as Tony Earley's dazzling *Mr. Tall* (2015), which, like *Suttree*, serves up tragicomic fantasies that at times nearly drown out the darker subtexts of isolation, abandonment, suicide, depression, and addiction—long running motifs of southern Appalachian experience.

As Edwin T. Arnold states, "McCarthy would certainly have heard versions of Jack Tales while growing up in eastern Tennessee; the most famous tellers of the Appalachian Jack stories lived in nearby western North Carolina and southwest Virginia, and the stories themselves traveled widely throughout the region." [24] Such tales ran in the McCarthy family, too, through Cormac's brother William Bernard "Bill" McCarthy, who staked his claim on them as an English professor at Pennsylvania State University's DuBois Campus and the books he wrote, including *The Ballad Matrix: Personality, Milieu, and the Oral Tradition* (1990). His deep interest in American Irish ballads and folklore led to his part in editing and compiling *Jack in Two Worlds: Contemporary North American Tales and Their Tellers* (1994) for the American Folklore Society following its 1987 meeting in Albuquerque. [25] The book traces the telling and individual accents in the tales into the second half of the twentieth century.

The Jack Tales offer the backdrop to another dynamite incident in the McCarthy canon from the pages of *The Orchard Keeper*, when Uncle Ather, the crusty old mountaineer, enthralls John Wesley and his friends with a tale of how he captured a baby "painter" (eastern mountain lion, which biologists now believe was not a distinct genus) with his bare hands, and has the scars to prove it: while on a blasting crew, accompanied by a crewman named (wink wink) Bill Munroe, he rescues a surviving "painter kit" from a blasted den which he gives to his wife, Ellen, who briefly domesticates before it disappears with mother-panther a few weeks later. [26]

Telling the story, the old man looks to the ceiling and the lantern light there—"the image of the lampflame on the ceiling, the split corona a doubling egg, like the parthenogenesis of primal light" (152). When he concludes his tale of the marauding and uncatchable she-panther returned for her kit, he says cryptically, "They's painters and they's painters. Some of em is jest that, and then others is right uncommon. That old she-painter, she never left a track one. She wasn't no common kind of painter" (157). The persistence of uncommon mountain panthers in the Appalachian Mountains finds verification in scattered newspaper accounts. For example, a panther reportedly resided near Madrid, Maine, a town at the northern end of the Appalachian range in 1908, prompting townspeople to cancel church socials and other entertainments and to commission a professional hunter. A female cat standing two and a half feet at the shoulder was tracked by a party of hunters "to a large cave, but none dared remain after dark for the panther to appear. [Joe] Dignard, who is an expert rifle shot, is armed with an automatic rifle, large revolver and hunting knife and has several sticks of dynamite with which to blow the cave and force the animal to fight in the open," the Waterbury *Evening Democrat* reported.[27] Closer to home, the *Nashville Union and American*, on May 15, 1874, reporting advocacy for a Knox County workhouse, also noted that a "wild cat was killed near Morristown [a town in the mountains east of Knoxville] a few days ago, the first one seen in that region in twenty years."[28]

Such was the world of *The Orchard Keeper*'s Uncle Ather, born into a bygone age. While relating his panther story, Ather's inner eye lingers on the long road behind and the short road ahead in his life, returning to memories of the loss of his wife. The circumstances surrounding the dissolution of Cormac's marriage in the early 1960s to his first wife Lee Holleman remain murky—according to Dianne Luce, researchers have yet to find an official record of the marriage, and it is not known how the separation unfolded—but since *The Orchard Keeper* was published in 1965, it seems possible that McCarthy was, with the split vision of a writer and his fiction, pondering his own sense of loss and guilt. Lee Holleman McCarthy was pregnant with their son, Cullen, by 1962, but by 1966 McCarthy was married to Anne DeLisle, until he left in 1976 without offering reasons, according to Anne.[29]

In any case, McCarthy reprises the image of the split globe of the lamp in *Suttree*, where it delineates differing moral destinies seemingly as arbitrary as the small space between a stillborn child and its living twin, a space as small but significant as the difference between natural evil (birth defects) and moral action, determined or otherwise (recall the discussion of chiral twinning from Chapter 2—a teratoma can be formed when a twin in utero kills its sibling and usurps its living matter, the embryonic version of Cain and Abel). In my reading, this parthenogenic line might be a touchstone for understanding some of McCarthy's relationships and decisions, and even the remarked-upon absent father phenomenon in many of his books. Significantly, parthenogenesis implies women without men and the secondary function of men. In his particular turn of phrase, Cormac candles the egg of creation itself, having spent his young life on the frontlines of TVA's systems of containment and eruption, living in a landscape where the age of combustion collided with the nuclear age, the topic of the engineering chapter that follows.

There are several implications to the convergence of the panther story and the author's broken union. First, primacy is given to the feminine in matters of creation and fertility. In the creative aspect, McCarthy points to the she-ness of godhead, not just as a nurturer and protector but in the initial bright light and parturition of creation, and the role of the feminine as both implacable destroyer (i.e., the mother cat who invisibly slaughters pigs and reclaims her young in the tale) and unacknowledged creator. The mother

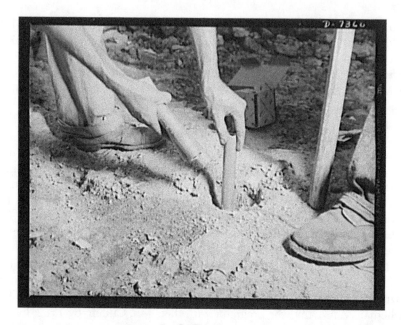

FIGURE 3.2 *Tamping a dynamite charge in bedrock at the site of the Tennessee Valley Authority's Douglas Dam on the French Broad River. Photograph by Alfred T. Palmer, June 1942. Library of Congress FSA/OWI Collection. SIS roll 32, frame 8. https://www.loc.gov/resource/fsa.8b04975/.*

of *The Road* cuts her own strings in this way. Second, and consistently in McCarthy's canon, the destructive (in this case, blasting) is conjoined with the mythically creative (the panther emerging from the ruins) in animistic and pre-Christian ways. Malkina cat-steps onto the page in *The Counselor's* later story as the embodiment of the life-force's amoral, insatiable hunger for more life, in which Westray, the male contributor, can be discarded after use.

In McCarthy's opus, then, dynamite is a force to set loose national and personal demons, to mark eras, to renegotiate power and the legacies of war, and to explore, through violent humor, the sublime forces of creation that hover on the edge of destruction. These forces that can lead to frozen accidents, and extinction without extinguishment, by extracting nonlinear patterns from simple rules.

*   *   *

Last, in McCarthy's expanding worlds, moral questions are never far away from complex events and chain of causation. Gene Harrogate's explosive misfortunes call to mind the totality of crime, punishment, and neglect leading up to his unholy yawp beneath the city. And cases like the Lebanon incident have been fertile grounds for legal tort debates through the ages. If the trailer was indeed labeled with appropriate warnings, would the construction company have been exonerated, partly or wholly? The ubiquity of dynamite and its poor regulation led to a swelling body of caselaw in the twentieth-century American South. For example, a similar set of circumstances played out in the adjoining state of North Carolina about fifty years earlier in which some lads, like the Lebanon boys, looking for something to shoot, chanced to fire upon a dilapidated shed in which the Norfolk & Southern Railroad Company stored a cache of dynamite (and left it untended for six months):

> Plaintiff had resided about 200 yards away from the shanty for about two weeks, but did not know that dynamite was in it. On Sunday morning, May 14, 1907, plaintiff, in company with McGhee, went to the river for the purpose of bathing. On their return they passed near the shanty, back of it near the river. McGhee had a pistol, and had fired four or five times at trees. He said that he had one more ball, and asked plaintiff to show him something to shoot at. While plaintiff was looking around, McGhee shot at the shanty. The ball passed through a hole and struck the dynamite, causing an explosion, blowing up the shanty, trees, etc., and injuring plaintiff. The shanty 70 yards away was injured, and several houses in Bridgeton shaken and window lights broken. The same effect was felt in Newbern. The plaintiff did not direct or advise his companion to shoot at the shanty.[30]

There was no warning sign that the shed was loaded with explosives, either. The North Carolina Supreme Court held that "as defendants owed no duty to anticipate injury caused by trespassers under such circumstances, the wrongful trespass of plaintiff's companion was the proximate cause of plaintiff's injuries, and not the storing of the dynamite, even if defendants were negligent in so storing the dynamite." The dissenting opinion offered a textbook example of the legal doctrine of *res ipsa loquitur* (the thing speaks for itself):

> The bare statement of the above facts is the statement of gross negligence. *Res ipsa loquitur*. Without such negligence on the part of the defendants, the plaintiff would not have been injured. McGhee would not have fired, if he had had any reason to suppose there was any dynamite stored in such close proximity to the public road and passing trains on the railroad.

When disasters unfold under complex circumstances, who bears the fault, especially when multiple parties are reckless? The case raised a familiar thicket of legal tort and philosophical questions regarding responsibility, agency, chain of causation, and contributory negligence, similar to the famed *Palsgraf v. Long Island Railroad Co.* (1928), which was adjudicated twenty years later. The celebrated case is such a standard example of tort law that some law schools turn it into a festive occasion by having a Palsgraf day, replete with mock legal duels between Justices Cardozo and Andrews. In *Palsgraf*, a man scrambling to catch a train was assisted onboard by railroad employees. In the chaos of the moment, he dropped his suitcase of fireworks, which detonated under the crush of the train wheels, causing a scale to fall upon and injure the plaintiff. The majority ruling, penned by the formidable justice Benjamin Cardozo (then serving on the New York Court of Appeals), ruled in essence that there was no liability to an unforeseeable plaintiff, so the railroad was off the hook for the damages (consistent with the NC Supreme Court ruling in the aforementioned *Fanning*).[31] Alternate potential defendants, as first-year American law students discuss with bright-eyed bewilderment, include the Suitcase Man, his mother, the fireworks and scale manufacturers, and perhaps God Herself.

Parsing the legal and ethical issues is more than an academic exercise where McCarthy's work is concerned. Underlying the legal questions are a series of unexplored metaphysical and scientific questions about time, space, probability, chance, not to mention the adequacy of physics as a descriptive system for parsing questions of causation. As chief counsel for the TVA, the senior Charles McCarthy was called upon routinely to dispose of matters of liability for the Authority's massive and diverse set of enterprises within a complex, multilayered legal landscape of federal, administrative, state, and local law. To that extent, Cormac was truly his father's exegete in fiction, lamping the subterranean dimensions of the philosophy of law and culpability in his own work, finally tunneling down to that recurrent place where human institutions struggle to redress injury and fault, whether in the form of a scofflaw like Sut Lovingood or a feckless ignoramus and human disruptor like Eugene Harrogate.

The origins of actions and their attribution are a matter of interest to philosophers and physicists alike, and it weaves ambiguously throughout McCarthy's works, including, notably, in the "flipism" questions of *No Country for Old Men*, a matter raised, fliply and memorably, by Donald Duck. Convinced by a crackpot professor to take all his decisions by a coin toss, Donald is nevertheless held strictly liable for the coin toss that sends him the wrong way on a one-way street. (The 50/50 coin flip remains a statistician's best friend, a redoubtable ally for benchmarking models in machine learning, counting large numbers, and optimizing results in Bernoulli's multi-armed bandits problem, according to data scientist Paul Tune.[32]) McCarthy's villainous Chigurh in *No Country for Old Men* insists on demonstrating the long chains of causation that in fact limit outcomes and lead us to our ends. South Texas scholar Linda Woodson wrestles with this problem in "'You Are the Battleground': Materiality, Moral Responsibility, and Determinism in *No Country for Old Men*," where she rejects vague notions of determinism and embraces the specificity of *causal determinism*, per John Martin Fischer and Mark Ravizza, as "The thesis that, for any given time a complete statement of the laws of nature, entails every truth as to what happens after that time." This sort of causal determinism, she observes, "continues to be a subject of debate among physicists and philosophers in the context of prominent physical theories, including classical mechanics, special relativistic physics, general relativity, and quantum mechanics."[33]

Moreover, "in theories of chaotic dynamical systems, it appears to be difficult to decide if randomness arises from genuine stochasticity or is governed by underlying deterministic laws."[34] This is precisely the area of inquiry trod by the Santa Fe Institute's merry band of complexity theorists, who are interested in sensitivity to initial conditions per chaos theory. Perhaps a butterfly's wing ultimately generated the larger gust but for which the Suitcase Man (whose identity in my mind blurs with the elusive luggage-toting Mustache Man of *Babar Loses His Crown*) would have held on to his case.

Woodson cites multiple examples of deterministic musing in McCarthy's oeuvre, including the words of the *ganadero* to Billy in *The Crossing*, "You do not know what things you set in motion. . . . No man can know. No prophet foresee. The consequences of an act are often quite different from what one would guess" (202). The Suitcase Man and his well-meaning attendants might rue the truth of this, as the rancher gestures toward the problem of sensitivity to initial conditions in complex systems—in so many words. So Chigurh might assert, with the benefit of hindsight, that "[t]he shape of your path was visible from the beginning," which makes his coin toss not an act of chance but revelation, consistent with the ancient world's view of the results not as determinative or chance-based, but merely revelatory of what the gods had decreed or fate had determined from the start.

And then Woodson comes to the crux of the matter as she sees it. She notes that Fischer and Ravizza "assert that while many contemporary physicists doubt that [strict determinism] is completely true, they do support the idea that although macroscopic events are not fully determined, they are very close to that."[35] The idea that macroscopic events are not fully determined contravenes Einstein's famous assertion that God does not play dice with the universe and the assumption of "no-spooky-action-at-a-distance," although many now believe that had Einstein lived longer he would have converged upon and accepted quantum mechanics, too.

Quantum indeterminacy might account for being "very close" but off the mark, and so, too, might be the failure of mathematical logic to fully realize, in Gödel's words, "a science prior to all others, which contains the ideas and principles underlying all sciences."[36] Aware of this history, in the final diptych of *The Passenger* and *Stella Maris* (2022), however, McCarthy acknowledges a set of attendant descriptive problems in mathematics as well, by harkening to the ways in which mathematics constantly undermines itself, from Emmanuel Kant's theories of intuition to Jules Henri Poincaré's philosophy of mathematics to Bertrand Russell's paradox to Kurt Gödel's "shocking incompleteness theorems." Those theorems, in Natalie Wolchover's words, "published when he was just 25 . . . proved that any set of axioms you could posit as a possible foundation for math will inevitably be incomplete; there will always be true facts about numbers that cannot be proved by those axioms. He also showed that no candidate set of axioms can ever prove its own consistency."[37] Ironically, Gödel's very name is eponymous with complex systems science, in a way. If the intellectual history of SFI were stitched into a volume, Dogulas Hofstadter's National Book Award and Pulitzer Prize–winning *Gödel, Escher, Bach: An Eternal Golden Braid* (1979)—his first book, published when he was thirty-five—would certainly reside in its Pentateuch for its early and celebrated excursions in complex systems, the looping nature of consciousness, and artificial intelligence. Martin Gardner (see Chapter 1), one of the finest scientific explainers of his generation, grasped the book's significance at once, writing in *Scientific American*, "Every few decades, an unknown author brings out a book of such depth, clarity, range, wit, beauty and originality that it is recognized at once as a major literary event."[38]

Not all physicists recognize Gödel's theorem as a bar to the long-sought Theory of Everything ("a hypothetical framework explaining all known physical phenomena in the universe"), but then, not all physicists affirm the value of a Theory of Everything or credit its attainability.[39] Einstein was flummoxed by the inability to reconcile gravity and electromagnetism; Stephen Hawking died without a redemption for quantum theory as he failed to find an answer for the paradox that black holes do not destroy information, and finally retrenched from his advocacy of a theory of everything in part because of Gödel (and while black holes cannot shrink per the calculations of general relativity, they can per quantum mechanics).[40] After a forty-year search, the existence of the Boson particle in was discovered in 2012, an event that McCarthy eagerly anticipated in a National Public Radio *Science Friday* interview the year before:

> What they're looking for, principally, at the Large Hadron Collider, I think, is the Higgs boson. And if they don't find that, there's going to have to be a lot of revision done because that so-called Higgs mechanism is what's responsible for supplying the masses to the particles in the standard model. And if they don't find some way to get these masses into the particles, they're going to have to do a lot of re-writing with physics for the last 40 years.[41]

Now, instead of validating the physics of Einstein, it might be the unraveling of standard theory. String theory has failed to bridge quantum and standard theory after thirty years of contortions, and, more recently, the discovery that the Higgs Boson particle is heavier than expected, if confirmed, threatens to unsettle the standard model of particle physics, regnant since 1973. "If the W boson is more massive than the Standard Model predicts," explains *The Economist*, "it implies that something else is tugging on it too—an as-yet undiscovered particle or force."[42] Every force evolves a form.

Moreover, emergent laws with their own logic govern the behavior of complex, evolving systems, and the work of SFI has been to probe them. For the Infinite Universe theory there could be a complementary infinite number of physical theories. Woodson places Fischer and Ravizza's notions of determinism against Carl Hoefer's observation, in the early 2000s, that "many physicists in the past 60 years or so have been convinced of determinism's falsity," basing their ideas on the concept that the Final Theory will be a variation of quantum mechanics, and that quantum physics by its nature rules out determinism. Regardless, as Hoefer concedes, arguments for and against determinism will continue the while, and thus, says Woodson, "The threat to moral responsibility remains the same in either case,"[43] and the question of moral responsibility even in a universe of causal determinism remains unresolved. John Grady Cole can admit to the judge that his failure to defend Blevins led to his death, and the events that followed, including the prison fight. All of these matters find taxonomic legal translation in terms of art such as proximate cause, duty to assist doctrines, degrees of murder and culpability, inchoate crimes, and the like—not coincidentally, areas of fascination for McCarthy. By way of application, consider these concepts by way of *No Country for Old Men* in such examples as Carla-Jean-the-Snitch (or at least one who goes back on her own word), Moss's return to the crime scene as water-bearer, the coin toss, Bell's sins of omission, and Chigurh's inchoate intentions in the motel closet.

They all point to problems of guilt and atonement, the impossibility of an ethics of pure reason, and the moral problems that make McCarthy's work at once vexingly contradictory and, finally, beautifully ethically opaque. Probability wishes to make a science of chance through measurement, yet entirely distinct schools of philosophic, scientific, and mathematical thought arise from the distinction (or non-distinction) between randomness and chance (a distinction to which the Santa Fe Institute owes its raison d'être).[44] McCarthy's intellectual journey, crudely described, might be ordered from *language* to *science* to *mathematics*, but instead of revealing the possibility of an ethical theory of everything (which in terms of physical theory remains a contested possibility) it may have proved the impossibility of his quest, and pointed, in a Kantian way, and consistent with Poincaré's philosophy of mathematics, to the value of intuition and the unconscious (cf., *The*

Passenger and *Stella Maris*, the Kekulé Essays, and the concluding chapter of this book).

The intrinsic human difficulty with distinguishing between chance, probability, and the random is encapsulated in a math puzzler recapitulated not long ago in *The New Yorker*:

> Winkler let loose with the last official mind bender, a gambling thought experiment involving a fictitious couple named Alice and Bob, who are famous in math circles. Each of them has a biased coin—fifty-one-per-cent chance of heads, forty-nine-per-cent chance of tails. They each start with a hundred dollars, flipping the coin and betting against the bank on the outcome. Alice calls heads every time; Bob calls tails. The puzzle: Given that they both go broke, which one is more likely to have gone broke first?
>
> Rosenthal looked thoughtful. "Every question that we were asked tonight," he said, "the answer is never what it seems."
>
> Most of the diners guessed Bob, but the correct answer was Alice. John Tierney, a former *Times* columnist and a math buff (he once wrote that recreational mathematics was "oxymoronic"), thought it over. "But, the longer Alice plays, the less likely she is to go broke," he said.[45]

In this trick question, everything hangs on the post hoc condition that "they both go broke."[46] Our confusion surrounding probability is likewise captured in the distinction between accidents, incidents, and coincidence (recall the line from *The Counselor* that cartel overlords have heard of coincidences but have never seen one, a shrewd position for the business-minded and the actuary). The poorly secured dynamite trailer in Lebanon was hardly an accident; the longer it remained in the field, the greater the chance of the intersection of its attractively volatile cargo with armed, rambling teenaged boys.

The *Tennessean* interviewed a "firearms expert" who drew distinctions between ammonium nitrate and nitroglycerin-based dynamite, parsed the exact mixture of the dynamite in Lebanon, and concluded that Pruett's rifle had "3 chances in 4 of setting off an explosion."[47] If true, the only "luck" the juveniles had was the improbable result that they took multiple shots before it detonated.

Such was the explosive state and State of violence of McCarthy's years as a young man in Tennessee, before and after his time in the air force. The wayward teens shooting at a trailer could only roll snake eyes so many times before their luck ran out. The Lebanon incident was just one of many that could have fired McCarthy's imagination, as the concentrated potential energy of East Tennessee dynamite transformed lives and landscapes and ricocheted its way into his prose and philosophy.

Notes

1   "Blast Rocks Lebanon, Demolishes Trailer," *Nashville Banner*, January 30, 1958, 2.

2   "Lebanon Blast Kills Youth," *Nashville Tennessean*, January 30, 1958.

3   "Blast Rocks Lebanon."

4   "State Officials Study Violations in Lebanon Blast," [Murfreesboro, TN] *Daily News Journal*, January 31, 1958, 1, 7.

5   "Boy Killed as Pal Shoots Into Dynamite," *Knoxville News-Sentinel*, January 30, 1958, 14.

6   Ibid.

7   "State Dynamite Law Century Old," *Nashville Banner*, January 30, 1958, 2.

8   Steven Johnson, *How We Got to Now: Six Innovations That Made the Modern World* (New York: Riverhead Books, 2014).

9   Michael Pollan, *The Omnivore's Dilemma: A Natural History of Four Meals* (New York: Penguin, 2006).

10 Patrick Benjamin Swart, "Blasting the Farm: Chemical High Explosives and the Rise of Industrial Agriculture, 1867–1930," PhD diss., University of Montana, 2014, 2.

11 Joseph Husband, *A Year in a Coal-Mine* (Boston: Houghton Mifflin, 1910), 17.

12 Bill McKibben, "In a World on Fire, Stop Burning Things," *New Yorker*, March 18, 2022, https://www.newyorker.com/news/essay/in-a-world-on-fire-stop-burning-things.

13 "Burglars Used Dynamite," *Atlanta Constitution*, September 21, 1900, 9.

14 "A Bank Safe Robbed," *Norfolk Landmark*, January 17, 1904.

15 "Primer Explosion Hurts Schoolboy," *Knoxville Journal*, February 7, 1952, 1.

16 "$85,000 Sought After Boy Hurt By Blasting Cap," *Knoxville Journal*, October 6, 1958, 15.

17 "Sevier Boy Loses Fingers in Dynamite Cap Blast," *Knoxville News-Sentinel*, December 17, 1959, 17.

18 "Man Hurt by Blasting Caps in Woods," *Knoxville News-Sentinel*, January 27, 1982, 6.

19 Ethan Siegel, "Sodium and Water React, and Quantum Physics Explains Why," *Big Think*, May 11, 2022, https://bigthink.com/starts-with-a-bang/sodium-and-water-react/.

20 George Dempster, "Like It or Not," *Knoxville Journal*, June 11, 1937, 6.

21 Jack Neely, *Knoxville's Secret History* (Knoxville, TN: Scruffy City Publishing, 1995), 46–7.

22 "Judge Nelson's Retirement Recalls How He Broke Up White Caps," *Knoxville News-Sentinel*, July 29, 1926, 8.

23   For an in-depth look at the incident, see Dianne Luce, "White Caps, Moral
      Judgment, and Law in *Child of God,* or, *The* 'Wrong Blood' in Community
      History," in *Cormac McCarthy: Uncharted Territories,* ed. Christine Chollier
      (France: Presses Universitaires de Reims, 2003), 43–59.

24   Edwin T. Arnold, "Cormac McCarthy's Frontier Humor," in *The Enduring
      Legacy of Old Southwest Humor,* ed. Edward Piacentino (Baton Rouge:
      Louisiana State University Press, 2006), 195. See also Sara Spurgeon,
      Exploding the Western: Myths of Empire on the Postmodern Frontier (College
      Station, TX: Texas A&M University Press), 2005.

25   William Bernard McCarthy, preface to *Jack in Two Worlds: Contemporary
      North American Tales and Their Tellers,* ed. William Bernard McCrathy,
      Cheryl Oxford, and Joseph Daniel Sobol (Chapel Hill, NC: University of
      North Carolina Press, 1994), ix–x.

26   Elsewhere, the blasted space of an old quarry takes on nearly sacred
      significance: "The tiered and graceless monoliths of rock alienated up out of
      the earth and blasted in ponderous symmetry, leaning, their fluted faces pale
      and recumbent among the trees, like old temple ruins" (189). Walking through
      the hallowed and abandoned site of southwestern progress, fugitive Uncle
      Ather follows the quarry road at night, "the limestone white against the dark
      earth, a populace of monstrous slugs dormant in a carbon forest," a scene as
      much at place in *The Road* as in McCarthy's first novel.

27   "Will Dynamite Panther," *Waterbury Evening Democrat,* April 27, 1908, 3.

28   "Tennessee News," *Nashville Union and American,* May 15, 1874.

29   Recorded interview with the author and Dianne Luce, Charlotte, NC,
      November 21, 2021.

30   Fanning v. J.G. White & Co., 148 N.C. 541 (1908).

31   Palsgraf v. Long Island Railroad Co., 248 N.Y. 339, 162 N.E. 99 (1928).

32   Paul Tune, "In Praise of the Coin Flip," *Towards Data Science,* January 19,
      2020, https://towardsdatascience.com/in-praise-of-the-coin-flip-238ddfb02cb9.

33   Laura Woodson, "'You are the Battleground': Materiality, Moral
      Responsibility, and Determinism in *No Country for Old Men,*" in *No Country
      for Old Men: From Novel to Film,* ed. Lynnea Chapman King, Rick Wallach,
      and Jim Welsh (Plymouth: Scarecrow Press, 2009), 2.

34   Ibid.

35   Ibid., 3.

36   James Gleick, *The Information: A History, a Theory, a Flood* (New York:
      Vintage, 1988), 181.

37   Natalie Wolchover, "How Gödel's Proof Works," Abstractions Blog, *Quanta
      Magazine,* July 14, 2020, https://www.quantamagazine.org/how-godels
      -incompleteness-theorems-work-20200714/.

38   James Somers, "The Man Who Would Teach Machines to Think," *The
      Atlantic,* November, 2013, https://www.theatlantic.com/magazine/archive
      /2013/11/the-man-who-would-teach-machines-to-think/309529/.

39 Adam Mann, "What Is the Theory of Everything," *Space.com*, August 29, 2019, https://www.space.com/theory-of-everything-definition.html.

40 Adrian Cho, "Hawking's Final Quest: Saving Quantum Theory from Black Holes," *Science*, March 20, 2019, https://www.science.org/content/article /hawking-s-final-quest-saving-quantum-theory-black-holes, doi: 10.1126/ science.aat6421.

41 Cormac McCarthy, "Connecting Science and Art," *Science Friday* interview by Ira Flatow, National Public Radio, April 8, 2011, https://www.npr.org/2011/04 /08/135241869/connecting-science-and-art.

42 Davide Castevecchi, "Particle's Surprise Mass Threatens to Upend the Standard Model," News, *Nature*, April 7, 2022, https://www.nature.com/ articles/d41586-022-01014-5; "A Hint of Excitement? Data Contradicting the Standard Model are Piling Up," *The Economist* 443, no. 9292 (April 16, 2022): 76–7.

43 Woodson, "You are the Battleground," 3.

44 For a complete discussion see "Chance versus Randomness," *Stanford Encyclopedia of Philosophy,* https://plato.stanford.edu/entries/chance -randomness/.

45 Dan Rockmore, "Dinner! Drinks! Denominators!," Puzzled, *New Yorker*, January 10, 2022, https://www.newyorker.com/magazine/2022/01/17/dinner -drinks-denominators.

46 I use this problem in the classroom when teaching *No Country for Old Men* to STEM majors. Applied to strategy it suggests that entering a game with advantage suggests a course of smaller stakes and endurance—call it Chigruh's Gambit/Moss's Folly. For variants and their mathematical proofs, see the answers to Joe P. Buhler and Tanya Khovanova's colorful Puzzles Column from the Mathematical Sciences Research Institute, https://www.msri.org/ system/cms/files/1033/files/original/2019-Fall-Emissary-Puzzle-Solutions.pdf.

47 "Rifle Had 3 Chances in 4 of Setting off Explosion," *Nashville Tennessean*, January 30, 1958, 2.

4

Engineering and the Built Environment

Hypanthropic Times— The Tennessee Valley Authority, Expanding Systems of Containment, and McCarthy's Imagination

There were no inviolate places, only outposts that were less visited than others. The Arctic was drilled for oil, great pools of waste oil seeping through glaciers. The continent was becoming Europe in my own lifetime and I felt desperate. The merest smell of profit would lead us to gut any beauty left, there was no sentimentality involved. We had been doing so since we got off the boat and nothing would stop us now (44).

–JIM HARRISON, *Wolf* (1971)

In some respects, the last chapter flows into this one, because we are not quite done with dynamite and its derivatives—a technology, certainly, but more, a technology that "has made almost every aspect of modern technology possible—and caused its share of disasters along the way."[1] As a facilitating technology it set the course for engineering what archeologist Christopher Witmore has called the Hypanthropic Age.[2] He coined the term to answer a need for describing human engineering on a geological scale, "apprehending and approaching that which is in excess of monstrosity," and as a new entry into the debate over whether the present age is the Holocene, Anthropocene, or something else.[3]

Certain facts from the article convey Witmore's concept tidily: the 4.3 billion tons of concrete that were poured in 2014 alone, the fact that an "internal combustion engine is born nearly as often as a human," or the observation that human activity in any given year now "shifts three times more earth than all world's rivers combined."[4] "As for 'water control and utilization' projects," Witmore explains, "more than one large dam has been built every day every year for the last sixty—the interruption of sediment flows has caused deltas to subside all over the planet." In a postmodern extension of TVA logic, China's Three Gorges Dam, symbol par excellence of human-constructed environmental engineering, measurably changed the earth's shape and rotation. Witmore concludes, along with many geologists, that humanity has become akin to a geological force "whose memories . . . will stand out in the 4.6 billion year archive of Earth history."[5]

Some part of the chemical signatures of the hypanthropic, petrochemical age will be found in the truly globe-spanning dispersal of products like plutonium, Styrofoam, and Teflon variants that will not degrade (hence the phrase, "Teflon is forever"). A recent study of the extremely remote and nearly uninhabited Cocos Islands published in *Scientific Reports* estimated that 262 tons of plastic rubbish had been deposited on its landmass, including "about 977,000 shoes." "Plastic production has increased dramatically over the last decade—in the last 13 years alone, we've manufactured nearly half of all the plastic produced in the last six decades," notes the paper, adding, "a recent global estimate finds that 5.25 trillion items of plastic are now in the ocean, which is more than the number of stars in the Milky Way."[6] Whether the hypanthropic plastic era will leave an enduring mark, a scrim of plastic in the geological record, is a matter of academic interest. We know that plastic is so ubiquitous and poisonous that it is raining from the sky, lodging in our lungs, entering plants and our bodies in the food we consume.

And we know that human-produced signs and wonders occasionally provide astonishing reminders of engineering's hypanthropic reach. An emblematic moment, relevant to the Louisiana lakes and salvage divers in the first part of McCarthy's *The Passenger* (2022), played out in 1980 when a Texaco oil drilling rig, the sort Llewellyn Moss might have crewed, was carrying out exploratory work on 1,300-acre Lake Peigneur. It accidentally struck a salt mine below because of a "mapping mistake" when "an engineer mistook transverse Mercator projection coordinates for UTM coordinates," affirming the old wisdom of the saying, *measure twice and cut once.* The "mistake" drained the entire 3.5 billion gallons of the lake in three hours, sucking in 11 barges and the rig itself. It also created, for a few days, a 160-foot waterfall, briefly the largest in the state, and permanently changed the depth of the lake and its water. As salt water backflowed from canals and forced air out of the mines, it created its own spontaneous *wasserlichtkonzerte* as 400-foot geysers erupted, too.[7] It is fair to say that the crew members who, miraculously, survived, wondered what man had wrought.

As Witmore puts it, "one thing is for certain: this 'anthropos' is unlike any *anthropos* known to our 'un-accelerated' forbears, those who understand humankind as mortals, subject to death, in contrast to the gods."[8] Dynamite has been supplanted by other tools and the splitting of the atom, which McCarthy marks rightly as the start of a new epoch in which humans would live under the shadow of their collective extinction, a definitive turning away from the unrelenting, entropic facts by which humans existence had been defined since the beginning of the species. In *The Passenger* Robert Western's father is a principal scientist in the Manhattan Project, one "who had created out of the absolute dust of the earth an evil sun by whose light men saw like some hideous adumbration of their own ends through cloth and flesh the bones in one another's bodies." Robert recalls that ultimately his father "spoke little to them of Trinity."

> Mostly he'd read it in the literature. Lying face down in the bunker. Their voices low in the darkness. Two. One. Zero. Then the sudden whited meridian. Out there the rocks dissolving into a slag that pooled over the melting sands of the desert. Small creatures crouched aghast in that sudden and unholy day and then were no more. What appeared to be some vast violetcolored creature rising up out of the earth where it had thought to sleep its deathless sleep and wait its hour of hours (368).

It is a scene that McCarthy reworks over and over again in his writing. In his opus, the end of the Neolithic might be marked on page 425 of *The Crossing.* Born in an era when cattle = chattel = wealth, Billy Parham, in common with all living things within a 160-mile radius from the Trinity Site, is awakened by the inhuman false sun and instant "noon" of the predawn atomic test at White Sands (July 16, 1945). Billy glimpses a "a broken rainbow or watergall" (a weather gall, from the Shakespearean *weathergall*) fading to a "dim neon bow." The scene marks a new age, fittingly, with a symbol of the covenant, broken, in which humanity has acquired godlike power, ushering in an "alien dusk and now an alien dark." With an explosive engineering feat, the ages themselves were riven; going forward, the chemical signature of the atomic and nuclear ages would reside in human flesh through bomb tests scattering isotopes into our bodies that previously had no part in the human fossil record. The novel's last line is a particularly bleak biblical and literary reference—the sun also rises, without distinction and indifferent to a human-induced apocalypse or hellscape.[9]

Consistent with the account from *The Crossing*, my colleague Dustin Benham had two great uncles who were doing farm work around Morton, Texas, some 210 miles as the crow flies from the Trinity test site, in the early hours of the July morning that ushered in an alien dark. According to the National Park Service, "the flash of light and shock wave made a vivid impression over an area with a radius of at least 160 miles."[10] Two of Benham's great-uncles witnessed what they described as a brief reverse sunrise, without noise or an aftershock—only bright light. Notwithstanding the official cover story of the day (that a large ammunition dump exploded), they were mystified and only later pieced together what had happened.

Larry Calloway gathered a number of additional eyewitness accounts, both unwitting and knowing, from various sources of that day,[11] among them ranchers and shepherds:

> John R. Lugo was flying a U.S. Navy transport at 10,000 feet 30 miles east of Albuquerque en route to the West Coast. "My first impression was, like, the sun was coming up in the south. What a ball of fire! It was so bright it lit up the cockpit of the plane." Lugo radioed Albuquerque. He got no explanation for the blast, but was told, "Don't fly south."

Witnesses in proximity described the light as brighter than lightning and the sun itself. Standing at roughly the same distance from the site as Billy Parham were Richard Harkey and his dad, Sparkey Harkey,[12]

> waiting in the dark for a train at Ancho station. "Everything suddenly got brighter than daylight. My dad thought for sure the steam locomotive had blown up." They were fifty miles and a mountain range away from Trinity Site. . . .
>
> Rowena Baca, among those interviewed by the Albuquerque Journal's Fritz Thompson on the 50th anniversary, remembered the red light reflecting off the walls and the ceiling in her grandparents' house at San Antonio, 35 miles northwest of Trinity Site. "My grandmother shoved me and my cousin under a bed," she remembered, "because she thought it was the end of the world."
>
> At daybreak, rancher Dolly Onsrud of Oscuro woke up and looked out her window and saw a strange cloud rising from the other side of the mountains—right about where her cattle-grazing land had been before the U.S. Army took it over three years earlier.

Billy Parham's location in *The Crossing* is more difficult to pinpoint, but there are scattered textual clues: he drifts east by horseback on the old road past the Santa Rita mines and on through San Lorenzo and the Black range before hitting blacktop. In those days before Interstate 25 overwrote the Camino Real, which blacktop he intersects is hard to ascertain (152? 27?), but fittingly, he is in the environs of the Jornada del Muerto desert, so named by Spanish explorers for the difficulty of the waterless crossing and frequent ambush point. The abandoned settlement where he spends the night (and encounters the ominous deformed dog) is likely on the fringe of the Black Range and probably lies in the current-day I-25 corridor, somewhere in the vicinity of Truth or Consequences, NM. Because McCarthy ground-truthed many of the locations in the Border Trilogy, the place might yet be pinpointed. Based on the textual clues, by my calculation, Parham can be confidently placed within around 60 miles of ground zero. Born into the last vestiges of the cowboy shepherd culture in 1928, anointed to adulthood by atomic fire, the epilogue of *Cities of the Plain* finds him at seventy-eight, full of years, without a home or friends, living under an overpass in Arizona, having played a cowboy as an extra in film.

Thus ends, argues Christopher Witmore, the Neolithic epoch,

> commonly understood as the earliest stage of agrarian society coming between nomadic forgers and configured societies (civilization). Among its features are a settled existence, living together in villages, agriculture— breaking earth on behalf of plants, sowing seed, encouraging crops, waiting for vegetal provisions to mature, bringing in the harvest, and protecting that investment—and a way of living with select animals, termed "domestication," specifically as partners in agriculture (co-workers, producers of fertilizer), as companions, as assets of wealth (livestock) and sustenance (milk, meat, warmth), for both agrarians and herders.[13]

The Border Trilogy is transfixed by the moment when the Neolithic order met the atomic age, a moment when global colonialism was energetically attempting to excise the vestiges of "incompatible" Neolithic cultures and folkways, much remarked in the various indigenous peoples of McCarthy's novels (and in *Blood Meridian*'s earlier premonitions).

More, I would add that a feature of the hypanthropic age, widely anticipated, is that technology raises engineering problems beyond the current grasp of engineers who fail to acknowledge complex systems. This is partly because emergent phenomena are not necessarily matched by emergent engineering—for example, warfare, global supply chain collapse, pandemics, social networks, and ecosystem collapse. Increasingly, artificial intelligence will be called upon to mend those gaps, raising another set of considerations that until recently were the stuff of science fiction, since machine learning and algorithm generation could generate a great deal of artificial stupidity as well. Consider, for example, the US energy grid, which was engineered based on projected growth with a number of reasonable, but faulty, assumptions.

FIGURE 4.1 *The Neolithic herder meets the Hypanthropic Petrochemical Age: Ches "The Goat Man" McCartney, immortalized just so by prose portrait in* Suttree. *"In just spring the goatman came over the bridge, a stout old man in overalls, long gray hair and beard. Sunday morning before anyone was about. A clicking of little cleft hooves on the concrete and the goats in their homemade harness drawing tandem carts cobbled up out of old signs and kindlingwood and topped with tattered canvas, horned goatskulls, biblical messages, the whole thing rattling along on elliptical wheels like a whimsical pulltoy for children. Loose goats flowed around the man and the wagon. A lantern swung from the hinder axletree and a small goat face peered from the tailboard, a young goat who is wearied and must ride. The goatman strode in his heavy shoes and raised his nose to test the air, the cart rumbled and clanked on its iron wheels and they entered the town." Cormac McCartney,* Suttree *(New York: Random House, 1979), 195. Anscochrome color slide of Ches McCartney taken by Eugène Rimar (1911–80) on Highway 19 somewhere in southern Georgia c. October 1958. This image is held in the private family collection of Peter Rima and licensed under the Creative Commons Attribution-Share Alike 4.0 International license.*

Global cryptocurrency mining, which uses more electricity than some of the world's countries,[14] contributes to climate change and has resulted in a race to the bottom for inexpensive energy, and, of late, renewable energy, drawing developers to Texas, where energy is inexpensive—and where the overtaxed electrical grid is dangerously vulnerable to collapse. It is rather astonishing to contemplate that the prize sought by the cryptological race is predetermined and finite, perhaps even a digital Ponzi scheme without "underlying assets or government backing,"[15] its value generated from the interaction of pure information with market forces that favor a certain type of desirable ephemerality. Infrastructure engineers of thirty years ago could scarcely have foreseen this element of bizarre and fluctuating architecture, and while scarcity-derived value can be seen as partly chance-based, we have the strange result of accelerating both climate change and black swan power events for the creation of an intangible, born-digital property.

Although the Santa Fe Institute is in no sense a wet lab, from its inception it has been invested in trying to understand emergent systems, with the result that it has often been in the vanguard of anticipating complex problems. And while its work is not applied at the Institute itself, it has started forging partnerships for applied complexity bringing "actionable insights" to projects that "occupy an often ill-explored, liminal space between the traditional domains of academic theory and application." Currently this takes the form of "two such projects; one on cyberhate, social polarization, and counterspeech, and a second on applied scale."[16] When artificial intelligence generates artificial stupidity with potentially catastrophic results, SFI takes an active interest in seeing the full dimensions of the problem. After all, its origin story is in the effort to fend off global destruction—led by scientists who made global destruction technologically possible to engineer, often through the specialized and terrifying knowledge that they cocreated. It is simplistic to say that SFI engages in reverse-engineering destruction through complexity theory, but it nevertheless might serve as a useful heuristic.

One possibility for SFI's future growth, then, is in laws that explain the limits of biology—again, part of its origin story and the SFI cohort's long-standing interest in scaling problems. In the best case, complexity science fosters the insights that reorient engineers, who must be fixated on the problems of the here-and-now, to make allowance for complexity and the emergent at a grand scale. Unlike Milan Kundera, I tend to think that the future is full of interest. Complexity science can help make future-approximations based on a crude look at the whole, even while acknowledging the vanity of its precise projection, rendering it, in that sense, "an apathetic void," as Kundera says.[17] However, I would cautiously point to a (non-causative?) inverse relationship between the level of aforementioned scientific mastery, as manifested in the energy currently devoted to information production, calculation, and data storage, and the reduction of biodiversity within complex earth systems.

Likewise, Edward Abbey wrote in his supercilious foreword (titled *Forward!*) to *Ecodefense: A Field Guide to Monkeywrenching* (1985),

With bulldozer, earth mover, chainsaw and dynamite the international timber, mining and beef industries are invading our public lands—property of all Americans—bashing their way into our forests, mountains and rangelands and looting them for everything they can get away with. This for the sake of short-term profits in corporate sector and multi-million dollar annual salaries for the three-piece-suited gangsters (M.B.A., Harvard, Yale, University of Tokyo, et alia) who control and manage these bandit enterprises.[18]

Abbey's early experiences, like McCarthy's, ran to Appalachia; he grew up in Indiana, Pennsylvania, a mining town. And his complaint might well have spoken to the son of an attorney whose stock in trade was defending the bulldozer, earth mover, chain saw, and dynamite on behalf of the TVA. While the TVA undeniably introduced certain modern amenities and sped the industrial economic development of southern Appalachia, readers of *Suttree* cannot help but wonder what Charles McCarthy Sr. might have made of its rebuke of progress.[19]

Moreover, Abbey's call to action before the gathering hypanthropic age is in its way a radical precursor to celebrated climate fiction (sometimes called cli-fi) works such as Kim Stanley Robinson's *The Ministry for the Future* (2020) or Barbara Kingsolver's *Flight Behaviour* (2012). It is also a reminder of Cormac McCarthy's personal esteem for Abbey and their shared views of an era of the natural world's ruination. "At the MacArthur [Grant] reunions," Richard B. Woodward wrote in 1992, "[McCarthy] spends his time with scientists, like the physicist Murray Gell-Mann and the whale biologist Roger Payne, rather than other writers. One of the few he acknowledges having known at all was the novelist and ecological crusader Edward Abbey. Shortly before Abbey's death in 1989, they discussed a covert operation to reintroduce the wolf to southern Arizona."[20] Before his passing in late 2020, Barry Lopez corroborated McCarthy's friendship with Abbey: "My understanding was that Cormack [*sic*] contacted Ed Abbey when I was writing *Of Wolves and Men*. A short while later Cormack contacted me and we went on to have many conversations over the phone about a range of things. I try to see him whenever I'm in Santa Fe."[21]

These three conservationist writers—Lopez, Abbey, and McCarthy—confronted mounting losses with a subversive solidarity. In 2014 the World Wildlife Fund warned that populations of the invertebrates they monitor had declined by half in the space of four decades.[22] A 2020 follow-up census was even more bleak: "Globally, monitored population sizes of mammals, fish, birds, reptiles, and amphibians have declined an average of 68% between 1970 and 2016." "The findings are clear," in the report's summary: "Our relationship with nature is broken."[23] And humankind is unambiguously to blame. Whether we are truly in the Anthropocene, or a sixth extinction, is an academic matter. By contrast, from 1974 to 1996—roughly the first two decades of the time period—the books and bound serials added to US academic libraries grew by about 80 percent (no correlation or causation inferred here—just a comparison).[24]

Instead of cycling into the biological complexity of earth systems, energy is being channeled into information capture and complex information systems; within fifty years, some have speculated that the number of digital bits may overtake the number of atoms on earth. At the fringe of theoretical physics, others, like Dr. Melvin Vopson at the University of Portsmouth and author of an unproven mass-energy-information equivalence principle, have suggested that "information is not just an abstract mathematical entity, but a 'physical, dominant, fifth state of matter.'"[25] (Contra Vopson, the fifth state is usually reserved for the Bose-Einstein condensate.) If correct, it would add a new layer of reality to McCarthy's suggestion in *The Road* that words (information) create reality. In the earlier *Whales and Men* (n.d.) manuscript, he floats the notion that language generates a property of secondary-ness that both confines us and allows the ratiocination of destruction: "Nomenclature is the very soul of secondhandness," Peter Gregory, an Irish aristocrat, says. "Language is a way of containing the world. A thing named becomes that named thing. It is under surveillance. We were put into a garden and we turned it into a detention center."[26]

Overmastering the natural world in our time is indeed a matter of containment and capturing information for simulation, which itself can be seen as a form of proxy nomenclature. As part of the bearing away of the world, I would submit that the current trend is to convert the living things of earth systems—vessels of complex information—into digital information, which from a detached perspective could also be seen as a highly engineered, mimetic simulation of living things. The great engineering project of the industrial revolution to the present day has been a quest for better fire, or, rather, its containment, to convert organic material in increasingly compressed forms (aka coal, dynamite, diamonds, enriched uranium, the atom itself) into energy and finally into the information signatures of the digital age. Now our energies are focused on generating the secondary, further confusing our sense of godlike mastery, in which we are self-appointed simulators rather than the simulated, and ecology itself is merely a system for our remote manipulation.

Presciently, in 2007, interviewer David Kushner watched as SFI president Geoffrey West took the stage to "suggest that there are two roads of thought we can choose to follow."

> The first is that we can believe in our power to heal the Earth before it's too late. He tells the audience that America needs to create a sort of reverse Manhattan Project—call it the Santa Fe Project—that that would bring together a consortium of top minds to engineer solutions to the escalating crises of sustainability and survival.
>
> Then again, he adds, there is another path. We can choose to believe that "we're a blip on the landscape, and that the vision of another SFI fellow, Cormac McCarthy, is where we're headed."[27]

Christopher Witmore asks, "Should we archaeologists posit here a close to ten millennia of agrarian styles of living for both humans and their fellow animals? Did we witness the end of the Neolithic in the mid-twentieth century?"[28] To this I would add Thomas Halliday's question, "What distinguishes us from other animals? There was no moment at which humanity suddenly arose."[29] If the atomic age marked the end of the Neolithic and the beginning of a special sort of post-human landscape, it coincides with the paradox that "exists at the heart of the notion of the Anthropocene as underlined by Slavoj Žižek: 'humanity became aware of its self-limitation as a species precisely when it became so strong that it influenced the balance of all life on earth.'"[30] As the current climate emergency demonstrates, we are on the cusp of extirpating many of the living things on the planet, including, perhaps, ourselves, precisely at the moment of our greatest "mastery."

Geoffrey West, the director of the Santa Fe Institute during much of McCarthy's time there, offered a very readable summation of one of SFI's keystone concepts in his book *Scale* (2017). West observes how, remarkably, everything from the bodies of animals to corporations are subject to certain scaling rules, and devotes a chapter to "The Fourth Dimension of Life: Growth, Aging, and Death," in which he takes up scaling as it pertains "to the sizes of animals, plants, and buildings" as a "template for considerations of limits to growth and sustainability."[31] Body mass, metabolism, and the scaling of circulatory systems explain why we don't have mammoth-sized ants, or vice versa, but even scaling can be warped by the pressures of a *hypanthropic* age. In yet another application of West's scaling rules, researchers recently determined that the

> median sizes of herbivores and omnivores have plummeted roughly 100-fold since the emergence of Neanderthals and Homo sapiens over the past few hundred thousand years . . . with the size of carnivores dropping by about 10 times in that same span. As a result, the U-curve that has persisted for so long has begun to noticeably flatten.[32]

Scaling rules still apply; what has changed is the human presence, exerting downward pressure like a giant flattening hand, such that the endemic *Pleistocene megafauna*—mammoths, giant camels, short-faced bears, dire wolves, the giant ground sloth—no longer roam the West Texas plains where I live, replaced by the monster trucks of *No Country for Old Men*.[33] In short, warping biological scaling laws is yet another sign of our outsize, hypanthropic influence on complex systems.

We know this, just as we know that the climate emergency demands urgent action, and that most people favor taking immediate action. So why the current stalemate? West theorizes that companies, for example,

> tend to become more and more unidimensional, driven partially by market forces but also by the inevitable ossification of the top-down administrative and bureaucratic needs perceived as necessary for

operating a traditional company in the modern era. Change, adaptation, and reinvention become increasingly difficult to effect, especially as the external socioeconomic clock is continually accelerating and conditions change at a faster and faster rate.

One might observe the same of academic institutions and systems of governance, which means that scaling up might introduce fragility just when collective action is required (cf., the results of the UN Climate Change Conference in Glasgow, aka COP 26).

By contrast, West notes that cities "become increasingly multidimensional as they grow in size. Indeed, in stark contrast to almost all companies, the diversity of cities, as measured by the number of different kinds of jobs and businesses that comprise their economic landscape, continually and systematically increases in a predictable way with increasing city size."[34] Cities, too, suffer from the effects of entropy, and notwithstanding the example of the Eternal City, collapsed cities dot the globe, since "the growth and mortality curves of companies closely resemble the corresponding growth and mortality curves of organisms."[35] However, West is gesturing specifically to the sources of urban resilience and how cities sustain themselves, asking, Is there a maximum size to cities? Or an optimum size? Living things and systems can become more organized and thus might be all that we know of negentropy—reverse entropy, in short—even if life is (as W. B. Yeats might say) "chained to a dying animal" and therefore subject ultimately to entropy.

Scaling can be connected to Robin Hanson's Great Filter Theory, an explanation for Fermi's Paradox (our failure to detect other intelligent life in the universe, or for it to colonize us): from prokaryotic life to complex life to a colonization explosion, this scaling up might naturally result in collapse, a sort of wall that organisms cannot surmount because of energy needs, overpopulation, and other factors. This seems rather anthropocentric and Earth-specific, but then, other commentators are pointing to the limitations imposed by the "laws" of ecology that impose additional filters on the here-and-now, including Rob Dunn in *A Natural History of the Future* (2022):

> Some of the laws of biological nature are laws of ecology. The most useful of these are universal. These biological laws of nature, like the laws of physics, allow us to make predictions. However, as physicists have pointed out, they are more limited than the laws of physics because they only apply to the tiny corner of the universe in which life is known to exist. Still, given that any story that involves us also involves life, they are universal relative to any world we might experience. Knowing about these laws helps us understand the future into which we are—arms flailing, coal burning, and full speed ahead—hurling ourselves.[36]

Dunn offers as an example a chilling experiment in which Harvard scientists deliberately exposed petri dish agar to increasingly high levels of antibiotic. "The experiment," Dunn writes, "mimicked the way we use antibiotics to control disease-causing bacteria in our bodies. It mimicked the way we use herbicides to control weeds in our lawn. It mimicked each of the ways we try to hold back nature each time it flows into our lives":

> The law of natural selection would predict that so long as genetic variation could emerge, via mutation, the bacteria should eventually be able to evolve resistance to the antibiotics. But it might take years or longer. It might take so long that the bacteria would run out of food before they evolved the ability to spread into the columns with antibiotics, the columns filled with wolves.
>
> It didn't take years. It took 10 or 12 days.[37]

So what? As Dunn points out, "An understanding of the law of natural selection is key to human health and well-being and, frankly, to the survival of our species." And he goes on to cite "other biological laws of nature with similar consequences," like the species-area law, which "governs how many species live on a particular island or habitat as a function of its size." (It turns out to be useful for predicting extinctions, too.) Then there is the law of corridors that "governs which species will move in the future as climate changes, and how." And the law of escape, which "describes the ways in which species thrive when they escape their pests and parasites"[38]—a topic of increasing relevance to humanity in time of climate change, if the speciesism can be forgiven, and one that complexity science is tooled to address. Perhaps its application can forestall the collapse of complex social and ecological systems, or at least try to contain hypanthropic floods and fires.

* * *

When Cormac's father, Charles McCarthy, Sr., was interviewed for the Tennessee Valley Authority's Oral History Collection on May 4, 1983, the elder McCarthy acknowledged that he "knew very little about TVA" before he took the position, and that he was primarily attracted to the salary, understandable in light of his expanding family.[39] Dennis McCarthy told me that his father might have considered a first position in California as well, which became a road-not-taken milestone in his life story. In any event, Charles Sr. moved his family to a stately home on Windgate Street in Knoxville's Sequoyah Hills neighborhood, the start of his journey to becoming chief counsel for the TVA.[40]

The portrait that emerges in the oral history interview is of a temperate, exacting, and cautious man, as befits an attorney-politician.

McCarthy served the public, the courts, and the politicians. He acknowledged that environmentalism "was not a big deal" at the time—though it would become a more acute issue for the TVA in the wake of the 1960s—and he expressed his admiration for Franklin Roosevelt's leadership and especially for the women he worked with, who at that time were few. In 1982 McCarthy published an article in *Tennessee Law Review*, in which he looked at the long history of efforts to curtail the powers of the TVA, arguing that it should continue to be "unshackled"—not legally constrained—in order to work effectively.[41]

So how is this relevant to his son's writing? For one thing, a significant portion of McCarthy's work is set against the positions his father took. *Suttree* is arguably McCarthy's most autobiographical work, and it concludes with its protagonist in flight from TVA improvements and a bridge to nowhere:

> Out across the land the lightwires and roadrails were going and the telephone lines with voices shuttling on like souls. Behind him the city lay smoking, the sad purlieus of the dead immured with the bones of friends and forebears. Off to the right side the white concrete of the expressway gleamed in the sun where the ramp curved out into empty air and hung truncate with iron rods bristling among the vectors of nowhere. (471)

The image is reprised in the ruined overpass of *The Road*, suggesting McCarthy's consistently skeptical view of the "good" accomplished by TVA's infrastructure and (social) engineers. If *Suttree* is largely an autobiographical reflection of years spent slumming in Knoxville's underworld, it is also a book set on the Tennessee River as it winds through Knoxville—the primary exhibit in the Tennessee Valley Authority's publicly proclaimed successes. The portrait offered by the son-of-lawyer McCarthy was far from flattering, however; Buddy Suttree's river flows with any imaginable sort of offal, and the novel begins with the dredging of a body from its waters. Later, a dead baby will drift by, and another corpse will be committed to it. Meanwhile, the poor and dispossessed, like the character Eugene Harrogate, are imprisoned or marginalized as squatters along its banks, while social reformers come across as ineffectual, bigoted zealots. Their campaigns of false baptism, false imprisonment, and false social reform run through the novel like its fetid river. And a significant undercurrent is the city's restlessness to discard its past along with the dead. The novel closes with the sort of large-scale engineering schemes that the TVA brought to Knoxville: "the destruction of McAnally Flats," a poverty-ridden section of town, as "yellow machines groaned over the landscape" (470). Throughout the novel the older orders and rhythms of southern culture are under threat from mechanized forces beyond, and so there is a trenchant critique of the world that the TVA brought to Knoxville.

Such a view is characteristically different from the one espoused by McCarthy's father, who admitted in a law review article, "Like the elephant in the fable of the blind men, TVA is all things to all people"—perhaps especially in Knoxville and its environs.[42] Though the article offers something of a march of statistics, the elder McCarthy would also write, in a passage that sounds like a promotional reel, "The effects of the program have been dramatic. Figures do not tell the story adequately. To understand what has happened, one has to drive through the Tennessee Valley and contrast its appearance today with its appearance 15 years ago."[43] Pointing to a common theme of mid-twentieth-century development in the American South, he notes that the "the hungry-looking mule is being replaced by modern machinery. The look of poverty has left the land."[44] Ironically, the whole of his son's novel was devoted to showing how the *look* of poverty might have left, but the fact of poverty remained—and one did not need to look far to find it.

Along these lines, the senior McCarthy goes on to tout successes in disease control: "The old muddy river has gone and with it have gone the annual threat of floods and the scourge of malaria. In its place are beautiful lakes which carry an increasing amount of commercial navigation and support a rapidly growing recreation and fishing industry."[45] In Cormac's *Suttree*, though, water-borne diseases are alive and well, including typhoid fever, not to mention non-waterborne ailments such as rabies and venereal disease. An atmosphere of wistfulness abides—for the lost civilization of trains, of Indians, of the simple agrarians and pioneers who built Knoxville. Yet the book is skeptical of any claim to human "civilization" in its obsessive return to the poor, the malnourished, the mentally ill, the addicted, the sick, the discarded, the imprisoned—all lamentable elements of society that run just under the surface. In other words, *Suttree* is devoted to exposing those very things that the law hopes to remedy, and that Cormac McCarthy's father had hoped to alleviate over the course of a career in public service.

At the same time, the young McCarthy's intellectual horizons and exposure to culture were almost certainly enhanced by his father's work. One finds a glimpse of this in the "old tattered barrister" whom Suttree meets, a man "who'd been chief counsel for Scopes, a friend of Darrow and Mencken and a lifelong friend of doomed defendants, causes lost, alone and friendless in a hundred courts" (366). The character in real life was John Randolph Neal, once a University of Tennessee law professor, supporter of the TVA, and, remarkably, a Scopes trial defender. The eccentric lawyer exhorts Suttree, "Follow the trade that you favor and you'll have no regrets in your old age" (367).

A world that accepts the purpose of engineering trade as human control of environment and containment per se (without regard to de-growth and scaling limitations) might be more susceptible to notions of autocracy and social engineering in the containment of what is deemed criminal as well as the meting out of punishment. The explosive energy of this relationship finds rich illustration across McCarthy's works (try applying it for yourself, and it soon becomes apparent that the containment motif runs through them all). Carceral systems of containment, leased convict labor, and energy extraction share a long history in the United States. By the time of Cormac's birth Tennessee was seeing a rise in dynamite tied to prison and labor incidents within the landscapes of southern Appalachia. In 1938, when Cormac was five years old, a group of prisoners working in the Brushy Mountain Prison coal mine blasted an exit from its clammy confines, leading to a short-lived exodus of thirty-eight of them. Twenty-one were quickly recaptured.[46] And on September 29, 1952, when McCarthy was a freshman at UTK before leaving for the air force, eight men were killed and twenty injured at the prison quarry near Knoxville after a blasting miscalculation. A county workhouse truck like the one that transports Buddy Suttree and Eugene Harrogate in *Suttree* had taken twenty-two prisoners to the site. In a detail that McCarthy might have appreciated, newspaper accounts stated a "quarry powder man" named Jess Loveday touched off the blast, by his account, using "fourteen cases and thirty sticks of dynamite."[47]

Strange to say, around twenty years later after the Brushy Mountain escape a group of convicts would combine the themes of both labor strife and prison rebellion in the coal mines. As late as 1960, three state-operated prison mines provided coal to be used in Tennessee state institutions. And as early as 1907, opponents objected to state management of mines on the grounds that they would generate larger profits if privatized. On July 13, 1959, ninety-five convict-miners captured three prison mine overseers and went on strike, threatening to "blow up the mine." The Assistant State Corrections Commissioner, A. Pat Patterson, brushed aside complaints about work conditions and substandard food, telling a reporter, "Some of the boys are just allergic to work." After three days the standoff ended as the prisoners succumbed to hunger—somewhat ironical given that two years earlier they had staged a hunger strike.[48] Even so, it seems they were capable of making good on their threat. Eleven bombs were rigged to explode if stepped upon, each fashioned out of two to ten sticks of dynamite, with fuses fashioned from the batteries of their miners' cap lights.[49]

The situation was reprised, again with three foremen taken as hostages, five years later in 1964. And once again, "the prisoners had dynamite and dynamite caps . . . available to them, but showed no disposition to use them," according to reportage. Apparently, the main check on impulse was self-preservation, an instinct that faltered from time to time. Asked what he thought would happen when he conferred with the prisoners during negotiations, the aforementioned Pat Patterson said, "You never know. You're dealing with convicts. They're not in here for being honest and telling the truth.[50]

Bear in mind that this "correctional" system, designed for self-sufficiency and the payment of social debts, in which coal was mined by prisoners to furnish electricity for state institutions, and which routinely loaded prisoners up with dynamite, is faithfully depicted in *Suttree*. As McCarthy hinted at the essential importance of his southern upbringing in an interview with David Kushner, "You grow up in the South," he said, "you're going to see violence.... And violence is pretty ugly."[51] Knoxville remained an erratically violent and unpredictable place, and the cast of misfits that McCarthy associated with during the *Suttree* years are representative of its undercurrents. One gets a sense of barely contained bedlam in an eyewitness Cormac McCarthy account he sent to Howard Woolmer in 1981:

> Went downtown yesterday & suddenly found myself surrounded by fleeing felons firing off revolvers, police running up the street returning the fire, several detectives dragging another man from an alley & fastening handcuffs & leg irons to him, etc. A newsboy rushed up to me with a microphone & a tape recorder & asked me what was going on. I told him I had no idea, but it certainly seemed like the good old days for a few minutes.[52]

The good old days never really went away in Knoxville as it combined volatile engineering projects with schemes to contain human vice. Homer Clonts, the city editor for the *News-Sentinel* whose excellent reportage of the Clinton riots was quoted earlier, was eating at the lunch counter of S.H. Kress when Robert D. Patty, a veteran of the Korean War, removed a lever action 30-30 from its box (purchased a week earlier at Sears after he lied about his mental health history), and started a shooting rampage outside Kress on May 11, 1976, killing three people among the lunchtime crowd on Gay Street.[53] One of the victims staggered into Kress, collapsed, and died.

The scene that Cormac encountered downtown in 1981 played out just a few blocks away from the Kress Building, following the escape of three prisoners from the Knox County Jail who were being transported to testify in the trial of a seventeen-year-old boy who had been raped while in custody at the jail. *Suttree*, too, hints at subtexts of sexual violence surrounding teenaged Harrogate and his harrowing trip through the carceral system. Edwin Arnold first pointed out the strong evidence that Lacey Rawlins is raped while incarcerated in Mexico, evidence that Jay Ellis codified in a note about the piece. The captain's interrogation of John Grady owes much to Hemingway's short story "A Simple Enquiry," in which a major makes his orderly deeply uncomfortable with his fastidious, gay mannerisms and his signaling homoerotic questions. In *All the Pretty Horses* John Grady Cole is flustered when the captain says of Blevins, "He don't have no feathers," indicating his personal knowledge that the prepubescent boy has no pubic hair (167). The same is said of Harrogate when he is deloused at

FIGURE 4.2 "*#9 gang on rock pile*" *(1975), JD Sloan. From the Library of Congress. While this prison work crew was probably in proximity to Nashville, it gives a sense of the contemporaneous carceral world of* Suttree.

the workhouse check-in (*Damn if you ain't barely feathered, said the man. How old are you?* [38]).

One might say that Cormac stepped squarely into his own fictional world, or vice versa, on that May day in 1981 as firing felons fled in the streets; but for the sexual abuse of a teenaged prisoner in Knoxville, the convicts would not have had a pretext for their escape. And twenty-six-year-old police officer Rick Grindstaff might not have been taken hostage at gunpoint "and forced to drive a commandeered car through a hail of police gunfire" before crashing into a bus. One of the fugitives, saying he was going to "blow somebody's head off," took a woman hostage in a nearby drugstore; police "hollered at her to dive" and "got the drop on him," according to Chief Detective Bill Wilson.[54] McCarthy's account jibes neatly with newspaper accounts, a familiar locale to him (including Gay Street, Main Avenue, and the Gateway Bookstore). The incident evokes the mayhem of Chigurh's drugstore layover in *No Country for Old Men*, another tale that blends together military-industrial violence and containment schemes.

In short, for Charles McCarthy's (and the TVA's) visions of legal and engineering containment to be realized it would be necessary to contain Knoxville's undesirables and poor, as well. The results are sometimes tragicomic: Harrogate's "beastiality" with a watermelon (vegiality, really) begins his captivity as a "white boychild" (139) unlikely to find an exit path from

Knoxville's carceral system. As with Blevins in *All the Pretty Horses*, his efforts to wrap dignity around himself cannot conceal a Huck Finnish past in which child abuse, violence, and neglect figure prominently (though in fairness he is selfish, feckless, and, as Suttree discovers, might be a truly hopeless case, incapable of sufficient self-reformation to receive aid from others).

Still, the comic relief in *Suttree* has a way of distracting the eye from the underlying realities of the bawdy house, the workhouse, the displaced, and the unhoused. Across novels, characters such as Blevins, Lacey Rawlins, and Harrogate suffer the common fate of the indigent, living in subterranean culverts, under bridges, the jailhouse. Harrogate the man-child takes his place among the men gathered in early hours around burning trashbarrels, alongside "black men in frayed and partly eaten greatcoats of their country's service" (438). He is sent up to Nashville on his first burglary charge, then graduates to his first stint at the storied Brushy Mountain State Penitentiary in Petros, a fact revealed when his half-sister searches for him, finds Suttree, and explains that Harrogate's mother has joined his father in death. McCarthy bathetically calls attention to Petros, Tennessee, its Greek roots as a stony place and the rock, the very cathedral, of the carceral kingdom. Prison walls, mountain faces, dams, and blast walls comprise various faces of local containments. Tennessee might truly be called the Petros State.

Suttree and Harrogate's time is served in a compound named for the Maloney family, whose patriarch James W. Maloney settled in the Maloney Heights Tract in the 1830s, and whose descendants included George L. Maloney, who fought for the Union during the American Civil War, and his bevy of descendants who created an investment group and with it an early suburban development. On the same grounds the Maloneyville Poor Farm rose, a county workhouse that by 1910 was home to 102 poor souls. Originally a poor farm or poor house replete with a potter's field (pauper cemetery) and chapel, the facility was quite suitable for a Country Mouse or even a City Mouse (as Harrogate is variously nicknamed), blurring the boundaries between the ancient almshouse, the so-called correctional facility, mental institution, and an unfree labor force. Public intoxication, petty crimes, pensioners, and debtors alike could be sent to the poor house where there were expected at minimum to provide the labor for their own upkeep, but more often, where they labored extensively, for example, in road construction, giving rise to the later "chain gang" road crew of pop culture imagination. Maloneyville was the generic local name for a cluster of enterprises—penal farm, workhouse, home for the destitute—and as *Suttree* documents, retained a good deal of its essential character as a hardscrabble, deeply racist, crackpot substitution for a social safety net.

In some respects, the system evolved from the exigencies of Reconstruction. "After the Civil War," writes David Sowell, under the crushing burden of state debt, Tennessee enacted a system of convict leasing to companies in order to raise much needed funds. Inmates were contracted out to mine coal and ore, repair or construct railroads, work in tanneries, and work on private farms. Conditions that convicts were forced to live under were brutal; often they were reduced to chattel in the records of the companies they worked for.[55]

Although the convict leasing system was formally abolished in 1893, its vestiges endured in the form of enforced labor schemes and in the gray areas of informal contracts that skirted the letter of the prohibition.

By the 1930s, the institution of the poor farm began to wane, at least in name, partly because of the introduction of federal social security legislation. Traditional American manual labor for engineering projects—navvies, chain gangs, and so on—yielded to the bulldozer the way mule-based agriculture gave way to the tractor in the postwar South. But change was slow to arrive in the rural South and its carceral systems, which constituted then and remain a perpetual backwater of bureaucratic and social neglect.[56]

Some of the change was nominal. In 1951 the Tennessee Reformatory for Boys was renamed the "State Training & Agricultural School."[57] The Maloney Poor House, named for its benefactor, was chastely rechristened The George Maloney Home in 1950s, repurposed as a tuberculosis sanatorium, and what might now be termed a community health clinic; eventually it became the site of Knox County's current Roger D. Wilson Detention Center. When he is slighted by his uncle for his time in "the goddamned penitentiary," Suttree returns the dig, "The workhouse, John. It's a little different. But I don't go around telling people that I've been in a T B sanitarium" (19).

Another aspect of the workhouse mission was to provide health care to assist in the containment of disease among the indigent and medically underserved. A 1944 Venereal Disease Information guide lists facilities for the treatment of syphilis and gonorrhea in Knox County at the County Jail on Gay Street (built in 1935), the County Workhouse Clinic in "Maloneyville," and the Juvenile Home Clinic on Middlebrook Pike.[58] A Juvenile and Domestic Relations Court had only recently been established. In the composite, these were the institutions Harrogate bounced through, described with all their rough edges, down to the dusting for "crotch crickets" that barely feathered Harrogate receives upon entry to the workhouse—a stint that he could easily have avoided by fleeing the hospital after he is peppered with birdshot in a melon patch—except that his pride could not bear the humiliation of wearing a hospital gown in the streets.

Harrogate likely misstates his age in a show of not-very-streetwise bravado, so that he is processed from the County Jail to the Maloney facilities (which is the site of the present-day Knox County Jail, also known as the Roger D. Wilson Detention Facility). In another comic misstep, when his jailers take pity on his tender age and put him to work in the kitchen,

FIGURE 4.3 *"Jailer Richard, county workhouse/jail, Nashville" (1975), JD Sloan.*
Library of Congress.

a trustee's most coveted job, he foolishly objects to the light duty, telling
Suttree he would rather swing a pick than "warshin dishes" (46). Harrogate
seems to be following the novel's character, a white man and "relic of the
Civil War" whose nickname is Nigger:

> I raised four boys and damned if they aint all in the penitentiary cept
> Ralph. Of course we all went to Jordonia. And they did have me up here
> in the workhouse one time but I slipped off. Old Blackburn was guard up
> there knowed me but he never would say nothin. Was you in Jordonia?
> Clarence says they aint nothin to it now. Boys, when I was in there it was
> rougher'n a old cob. Course they didnt send ye there for singin in a choir.
> I done three year for stealin. Tried to get sent to T S I where they learn
> ye a trade but you had to be tardy to get in down there and they said I
> wasnt tardy. I was eighteen when I come out of Jordonia and that was in
> nineteen and sixteen. (25)

The old man references Nashville's Tennessee Reformatory for Boys
("Jordonia"). Either he or the text mistranscribes the initials of the
Tennessee Industrial School (T S I), which opened, along with the Tennessee
State Training and Industrial School for Colored Boys in Pikeville, in 1911.
Jordonia was renamed the State Vocation School for White Boys in 1955,
at the height of integration backlash, and renamed again as the Spencer
Youth Center in 1971, and again in 1990 when it became "Woodland
Hills Youth Development Center."[59] When the State of Tennessee abolished

corporal punishment in juvenile institutions in 1978, the notorious solitary confinement cells at Jordonia were closed,[60] yet Woodland Hills commanded the national media spotlight in 2014 when thirty-two teens rioted, "overpowered security personnel and kicked through a dorm wall" before escaping under a fence, challenging, as reported by the *Tennessean*, "the image of a reform school and paint[ing] more the image of a jail facility for youths."[61] Consequently, the governor of Tennessee called for the institution's operation to be outsourced, in keeping with the national trend toward privatization of incarceration. In 2018, the security was downgraded, the building renovated and leased to Florida-based Truecore Behavioral Solutions, and rechristened with yet another Orwellian title, Gateway to Independence, still in operation as of this writing.

So runs the strange career of Jordonia, in the burgeoning area of "behavioral health" and human containment some 110 years after its founding.[62] The point here is that social engineering schemes run with the built environment, are embedded in ways variously calculated to promote social invisibility and to give visibility to punishment, and no nomenclature has freed us from this fact.

*   *   *

In late 2015 I recorded an interview with Mary Jane Salyers (b. 1934), author of *Appalachian Daughter*, who grew up on a farm along a rural route about three miles west of Petros—a place name that lies thickly on Harrogate's half-sister's tongue in *Suttree*—and the Brushy Mountain Prison during the 1940s and 1950s. I wanted to see if she had any memories to match McCarthy's depiction of incarceration and local culture in East Tennessee, especially since her family's farm was virtually adjacent to the outlying prison work farm at Brushy Mountain. She recalled how her mother

> was standing at the kitchen sink and this man came to the window that was above the sink—in the summertime the window was open—and he asked her if he could have a drink of water. And so there was a rock wall out the back door and she took him up a glass of water out and he asked if it was okay if he set there and she told him yes and he said I'm helping train the dogs. I am trying to evade them and they're training them to track me and he said I was in the creek and my cigarette papers are all wet. If I could just have a paper sack I could tear off a piece of cigarette paper. Well, he set there and smoked his cigarette and then he asked mother if it would be all right if he would go down to the edge of the woods down over the hill and there was some kind of little building there. He said could he climb up on top of that building. She told him yes and so he left, and that all we saw of him.

> And then my father came home and she told him. He said, "Well, did you call the prison and ask them if they had people out training dogs and if this was a legitimate story?" She said, "I didn't think of that," so he called and they said, "Yeah, we've got a crew out playing and training the dog."[63]

In its depiction in media and films such as *Cool Hand Luke*, and before them, in slave narratives, the bloodhound has become a national symbol of containment systems in the American South. While it is not entirely clear if the "construction gang" at the close of *Suttree* is a chain gang comprised entirely of prisoners, the "enormous lank hound" that "sniff[s] at the spot where Suttree had stood" as he hitchhikes away from Knoxville (and watches men beg for a drink of water in a striking coincidence with Salyers's story) offers a clue (470-1). "Somewhere in the gray wood by the river is the huntsman and in the brooming corn and in the castellated press of cities. His work lies all wheres and his hounds tire not," McCarthy writes. Perhaps this is the ancient pattern of the world, yielding to the hypanthropic age; it is, in any case, a story about the timelessly yoked campaigns to control at once the disempowered and the environment. Of the hounds, Suttree says, "Fly them," as he flees Knoxville (471).

Mary Jane Salyers indicated that fugitives were common enough, and that some local families even left basic supplies out in case they happened by, whether as an act of solidarity or to keep them from invading their homes.

> We were so close to the prison that if anybody was trying to escape from there they wouldn't be fooling around, they'd be getting out of the county as fast as they could, or they'd be sleeping in your barn to get ready for their next move. There was once when my father went into the church, and they had these heavy fabric curtains that they used to block off Sunday school rooms. Somebody had taken one of those down and it was all lying down and they figured that somebody had come in and taken the curtain down as a blanket to curl up in and get warm. The guess was it probably was [a fugitive].

Although convict labor schemes had long since been formally outlawed, she recalled reading "about a dispute between the miners who were unionized and they were sending some prisoners to work in the mine—not Brushy Mountain—it was that little town that was over close to where they later made Norris Lake—and that there was a riot that happened." Salyers was likely thinking of the Coal Creek War of the early 1890s. "Occasionally," she said, "when there was a grave needing to be dug, they would ask the prison to send some prisoners to help dig the grave and they would do that."

I shared with her a passage from *Suttree* in which Cornelius Suttree's imagination is sparked by a makeshift bar tabletop fashioned from the smooth side of a salvaged gravestone. The scene depicts a family evicted by TVA development, in terms that seem a nod to William Faulkner's *As I Lay Dying,* in which a family searches for dry ground for its dead:

> Suttree traced with one hand dim names beneath the table stone. Salvaged from the weathers. Whole families evicted from their graves downriver by the damming of the waters. Hegiras to high ground, carts piled with battered cookware, mattresses, small children. The father drives the cart, the dog runs after. Strapped to the tailboard the rotting boxes stained with earth that hold the bones of the elders. Their names and dates in chalk on the wormscored wood. A dry dust sifts from the seams in the boards as they jostle up the road. (113)

The passage prompted Salyers to recall a critical chapter in her family's history:

> In my father's family there were two of his siblings who died as children, and they were buried in the area where they lived at that time and then my grandfather moved the family to the Knoxville area. And then when . . . the TVA came and they started flooding, they did move the cemeteries, they did move the bodies, and I have been to see the graves of these two little children in their new location because where they were first buried is underwater. And I have heard people say that there were homeplaces were covered with water where there had been a fire in the fireplace for over a hundred years without it going out. They were pretty particular about not letting the fire go out. So, my mother's father, his family owned a big farm—the Norris Lake is on I believe it's the Clinch River—and this farm was down by the Tennessee River. I can remember as a very small child that we went to visit my grandfather and there was an island out in the Tennessee River. There was this little ferry that was just big enough to drive one vehicle on and then they pulled it across to this island you could actually drive the car then and my grandfather was farming that little island. And when they built the Watts Bar Dam the island was there no more.[64]

Norris Lake is a point of memory for McCarthy, too. The father of *The Road,* whose reminiscences verge on a roman à clef for some of Cormac's life experiences, recalls watching a falcon at Norris Dam (17), and, as Wesley Morgan documents, the lake that the boy and his father visit after descending from Cumberland Gap is indeed Norris Lake.[65] In any case, in Mary Jane Salyers's words we hear the experiences of the displaced existing on marginal land, and more, a source for McCarthy's motif of variously keeping and "carrying the fire," in a catchphrase that has become the unofficial motto of his devoted readers. Only later did I appreciate the full

FIGURE 4.4 *From a miscellaneous lot of photographs by Barbara Wright, including images of Tennessee, and the TVA, Knoxville, n.d., Library of Congress. Although the LOC dates the image "ca. 1941," the woman has been identified as Swiss writer Annemarie Schwarzenbach, who was traveling across the United States with Barbara Hamilton-Wright in 1936–7. Cf., https://gtmuseum.org/pages-from -the-photography-collection/iv-annemarie-schwarzenbach/. The photo gives a sense of the Tennessee River crossing as remembered by Salyers before the Watts Bar Dam had been constructed.*

import of Salyers's anecdote. McCarthy's fascination with Irish and Scots Irish folkways has been much remarked in the critical literature surrounding his work. While many cultures observe the sanctity of the hearth (see for example the particularly formulaic and ancient construction of Mayan hearths in Yucatan), the Irish and Scots Irish ascribe profound meaning to it, as E. Estyn Evans explained in *Irish Folk Ways*:

> The kitchen and the hearth are the very core of the Irish house, and the turf fire burning continuously day and night, throughout the year, is the symbol both of family continuity and of hospitality towards the stranger.

When it goes out, it has been said, the soul goes out of the people of the house. The fire serves not only to prepare food and dry clothes, to bring warmth and comfort to the family and to ailing animals, but also to keep the thatch dry and preserve the roof timbers, so that "when the smoke dies out of a house, it does soon be falling down."[66]

Estyn Evans notes the ancient use of fire as a ward against evil spirits, and that "soot carried in the pocket, for example, gave protection on a journey"—a symbolic way to *carry the fire* (*The Road*), a catchphrase and imageset so central as to become shorthand for McCarthy's works in the way of "broken places" in Hemingway's. Moreover, fire helped to define gender roles in Irish culture, from the arrangement of people around a hearth to its preservation:

The fire is, with her children, the "care" of the woman of the house. "A woman can never go to the fire," runs an Irish saying, "without tampering with it." The manifold duties of the housewife keep her moving about the hearth, tethered to it for most of the day. Her last duty at night is to "smoor" it, burying a live turf in the ashes to retain a spark which can be fanned into a blaze next morning. This custom was fortified by the belief that the good folk, the fairies, would be displeased if there were no fire for them through the night. I have sat at fires which, it was claimed, had not been allowed to go out for over a century, or for as far back as family memory could go. It is to the fireside seat that the visitor is invited, for this is the place of honour, and it is around the fire that tales of old time are told. The magic of the open fire, playing upon the fancies of generations who have gathered round it, has engendered a host of beliefs and portents. The fire can give warning of wind and weather, of lucky and unlucky visitors, of marriage and death. Above all, it is a shrine to which ancestral spirits return, a link with the living past.[67]

Suttree asks what happens to the spirits under the waters, in the necropolis beneath Knoxville, in the bones scraped up along the new highways, and the ghosts that wash over Gene Harrogate during the bank-blasting incident. Fly them? Underpinning this essential question are deep folkways concerning honoring the dead, spirits, and the living. When Mary Jane Salyers described the extinguished hearth, she was tapping into a deep cultural tradition, remarking a cultural desecration and its finality. Today it is hard to imagine a multigenerational American home, continuously occupied for a century or more, and in some respects *Suttree* is a paean to the passing of that epoch. Suttree, too, proves an astute observer of Knoxville's structural and social engineering, employing an ancient vocabulary that joins pre-Christian "superstition," the sacred, and the profane.

In this ancient architecture—the meaning under Mary Jane Salyers' story of life in an age of limited combustion encountering a hypanthropic and indeed atomic era—we are reminded that ritual and religion purposefully passed down formulas for survival. Reflecting on the tautology of science that ultimately can only draw conclusions within the terms it sets, Timothy Andersen writes,

> Language and mathematics are a means of controlling and modifying collective human action so that work gets done.
>
> This is language as culture rather than language as picture. And culture includes ritual. Like all ritualistic communities, physics contains its rules, interpretations, specialised vocabulary, a community of adherents who are admitted to the arcane arts, levels of indoctrination, and gatekeepers. . . . Humans have spent hundreds of thousands of years navigating a hostile planet by encoding information crucial for survival into ritual, which can then be transmitted across generations. When we invented the scientific method only a few hundred years ago, we had to graft it onto that part of our nature in order to pass it down the generations, hijacking an ancient and effective cultural mechanism for a new purpose.[68]

"Stacking up stone is the oldest trade there is," McCarthy told Woodward in 1992. "Not even prostitution can come close to its antiquity. It's older than anything, older than fire. And in the last 50 years, with hydraulic cement, it's vanishing. I find that rather interesting."[69] When Suttree endures his hallucinatory trial, as part of his vision of hell unleashed, he sees "simmering sinners with their cloaks smoking carry the Logos itself from the tabernacle and bear it through the streets while the absolute prebarbaric mathematick of the western world howls them down and shrouds their ragged biblical forms in oblivion" (458).

Mathematic, as distinct from mathematics, was understood in Middle English to mean something more like philosophy—a single, unified science, in an endless conversation about limitations, containment, and humanity. Whether it shall hijack the logos or reveal it will have a profound bearing on the human-engineered climate emergency and how we follow, or are driven down, the road ahead.

Notes

1   Kate Morgan, "How Dynamite Shaped the World," *Popular Mechanics*, May 26, 2020, https://www.popularmechanics.com/science/a32447280/history-dynamite/.

2   Christopher Witmore, "Hypanthropos: On Apprehending and Approaching That Which is in Excess of Monstrosity, with Special Consideration Given

to the Photography of Edward Burtynsky," *Journal of Contemporary Archaeology* 6, no. 1 (June 2019): 136–53, doi: 10.1558/jca.33819.

3 Regarding this debate, see Giacomo Certini and Riccardo Scalenghe, "Holocene as Anthropocene," *Science* 349, no. 6245 (July 17, 2015): 246, DOI: 10.1126/science.349.6245.246-a.

4 Witmore, "Hypanthropos," 144.

5 Ibid., 139.

6 Yasemin Saplakoglu, "Scientists Went to One of the World's Most Remote Island Atolls. They Found 414 Million Pieces of Plastic," *Live Science*, May 20, 2019, https://www.livescience.com/65520-plastic-pollution-cocos-islands.html.

7 Ken Jennings, "How Louisiana's Lake Peigneur Became 200 Feet Deep in an Instant," *Condé Nast Traveler*, July 11, 2016, https://www.cntraveler.com/ stories/2016-07-11/how-louisianas-lake-peigneur-became-200-feet-deep-in-an -instant. Even after this incident, the salt dome under the lake is being used to store natural gas.

8 Witmore, "Hypanthropos," 139.

9 Quite a few articles concern this scene and scrutinize various encounters with dogs at the novel's end. Cf., Alex Hunt, "Right and False Suns: Cormac McCarthy's *The Crossing* and the Advent of the Atomic Age," *Southwestern American Literature* 23, no. 2 (1998): 31–7; Petra Mundik, "The Right and Godmade Sun: Fate, Death, and Salvation in Cormac McCarthy's *The Crossing*: Book IV," *Southwestern American Literature* 37, no. 2 (Spring 2012): 10–37; John Wegner, "'Wars and Rumors of Wars' in Cormac McCarthy's Border Trilogy," in *A Cormac McCarthy Companion: The Border Trilogy*, ed. Edwin T Arnold and Dianne C Luce (Jackson: University Press of Mississippi, 2001), 73– 91; Yujing Sun and Junwu Tian, "An Outcast in an Alien Land: The Metaphor of Dogs in Cormac McCarthy's *The Crossing*," *The Explicator* 79, no. 1–2 (2021): 91–6; Robert H. Brinkmeyer Jr., "A Long View of History: Cormac McCarthy's Gothic Vision," in *The Palgrave Handbook of the Southern Gothic*, ed. Susan Castillo Street and Charles L. Crow (London: Palgrave Macmillan, 2016). Robert H. Brinkmeyer connects the dog that appears at the end of the novel ("the harbinger of God knew what") to the "atomic bomb's terrifying power to disfigure and destroy" (179).

10 National Park Service, "Trinity Site," White Sands National Park, September 18, 2017, https://www.nps.gov/whsa/learn/historyculture/trinity-site.htm.

11 Larry Calloway's fiftieth-anniversary retrospective originally ran on the front page of the Albuquerque *Sunday Journal* ("The Nuclear Age's Blinding Dawn," July 9, 1995) and was deemed significant enough to be printed in the Senate Congressional Record: https://www.congress.gov/104/crec/1995/07/17/CREC -1995-07-17.pdf. The excerpts here are from a rewrite of the original article, published on his personal website, now archived by the Internet Archive: https://web.archive.org/web/20051018112209/http://larrycalloway.com/ historic.html?_recordnum=105.

12   No, I am not making this name up.

13   Christopher Witmore, "The End of the 'Neolithic'? At the Emergence of the
     Anthropocene," in *Multispecies Archaeology*, ed. Suzanne E. Pilaar Birch
     (London: Routledge, 2018), 26.

14   Jon Huang, Claire O'Neill, and Hiroko Tabuchi, "Bitcoin Uses More
     Electricity Than Many Countries. How Is That Possible?" *New York Times*,
     September 3, 2021, https://www.nytimes.com/interactive/2021/09/03/climate/
     bitcoin-carbon-footprint-electricity.html.

15   Mitchell Zuckoff, "Is Cryptocurrency a Ponzi scheme?," *Boston Globe*, May
     24, 2022, https://www.bostonglobe.com/2022/05/24/opinion/is-cryptocurrency
     -ponzi-scheme/

16   "Applied Complexity Projects," Office of Applied Complexity, Santa Fe
     Institute, May 27, 2022, https://www.santafe.edu/applied-complexity/fellows.

17   Milan Kundera, *The Book of Laughter and Forgetting* (New York: Alfred
     A. Knopf, 1980), 22. McCarthy once wrote to Howard Woolmer, "I read
     Kundera's book on the novel yesterday. Refreshing to find someone almost as
     pessimistic as myself. Or nearly so." Quoted in Michael Lynn Crews, *Books
     Are Made of Books* (Austin, TX: University of Texas Press, 2017), 196–8.

18   Edward Abbey, "Forward!," foreword to *Ecodefense: A Field Guide to
     Monkeywrenching*, ed. Bill Haywood and Dave Foreman (Tucson, AZ: Abbzug
     Press, 1993), 3–4.

19   According to its 2021 strategic report, today the TVA "is the largest public
     power provider in the United States and the third-largest electricity generator
     in the nation," supplying energy to "approximately 10 million people and
     over 750,000 businesses across the seven states of the Tennessee Valley
     region."

20   Richard B. Woodward, "Cormac McCarthy's Venomous Fiction," *New York
     Times Magazine*, April 19, 1992, 28–31.

21   Barry Lopez, personal communication to the author, February 10, 2020.

22   Damian Carrington, "Earth has Lost Half of Its Wildlife in the Past 40 Years,
     Says WWF," *The Guardian*, September 30, 2013, https://www.theguardian
     .com/environment/2014/sep/29/earth-lost-50-wildlife-in-40-years-wwf.

23   World Wildlife Foundation, *Living Planet Report 2020 - Bending the Curve of
     Biodiversity Loss*, ed. R.E.A. Almond, M. Grooten M. and T. Petersen (Gland,
     Switzerland), https://livingplanet.panda.org/en-us/.

24   Margaret Werner Cahalan, *The Status of Academic Libraries in the United
     States: Results from the 1996 Academic Library Survey with Historical
     Comparisons (Survey Report)* (U.S. Department of Education, Office of
     Educational Research and Improvement, 2001).

25   Manasee Wagh, "Information Could Be the Fifth State of Matter, Proving
     We Live in a Simulation," *Popular Mechanics*, March 20, 2022, https://www
     .popularmechanics.com/technology/a39588076/information-could-be-the-fifth
     -state-of-matter/.

26   "Whales and Men" [screenplay], n.d., final draft, printout with no corrections, Cormac McCarthy Papers, box 97, folder 5, 57.

27   David Kushner, "Cormac McCarthy's Apocalypse," *Rolling Stone,* December 27, 2007, 43–53, http://www.davidkushner.com/article/cormac-mccarthys -apocalypse/.

28   Witmore, "The End of the 'Neolithic'?," 40.

29   Thomas Halliday, "Portrait of the Human as a Young Hominin," *Nautilus,* May 25, 2022, https://nautil.us/portrait-of-the-human-as-a-young-hominin-18441/.

30   Slavoj Žižek, "Ecology Against Mother Nature: Slavoj Žižek on Molecular Red," *Verso Books Blog,* May 26, 2019, quoted in Witmore, "Hypanthropos,"142.

31   Geoffrey B. West, *Scale: The Universal Laws of Growth, Innovation, Sustainability, and the Pace of Life in Organisms, Cities, Economies, and Companies* (New York: Penguin Press, 2017), 158.

32   The U-curve is the shape that results when "graphing the diet-size relationship of terrestrial mammals . . . when aligning those mammals on a plant-to-protein gradient." Scott Schrage, "Human Disrupting 66-Million-Year-Old Feature of Ecosystems," University of Nebraska-Lincoln Institute of Agriculture and Natural Resources News, April 25, 2022, https://ianrnews.unl.edu/humans -disrupting-66-million-year-old-feature-ecosystems.

33   Ibid.

34   West, *Scale,* 32.

35   Ibid.

36   Rob Dunn, "Natural History, Not Technology, Will Dictate Our Destiny," *Wired,* January 6, 2022, https://www.wired.com/story/natural-history-will -dictate-destiny/. Dunn's book from which the article is condensed is *A Natural History of the Future: What the Laws of Biology Tell Us about the Destiny of the Human Species* (Basic Books, 2021).

37   Ibid.

38   Ibid.

39   McCarthy, Charles J. Oral history interview by Mark Winter. May, 1983. Transcript, TVA Employee Series, Tennessee Valley Oral History Collection, TVA Library. Knoxville, TN.

40   John Shearer, "Cormac McCarthy's father was his first boss at TVA," *Knoxville News Sentinel,* May 8, 2019, https://www.knoxnews.com/story/ shopper-news/bearden/2019/05/08/cormac-mccarthys-father-charlie-van-bekes -first-boss-tva/3586351002/.

41   McCarthy, Oral history interview.

42   Charles J. McCarthy, "The Tennessee Valley," *Town Planning Review* 21, no. 2 (July 1950): 116.

43   Ibid., 125.

44   Ibid., 128–9.

45 Ibid., 128.

46 "38 Dynamite Way from Prison Mine," *Atlanta Constitution,* March 28, 1938, 14.

47 "Dynamite Blast Kills 8, Injures 20 At Prison Quarry," *Washington Post,* September 30, 1952, 9.

48 "Convicts in Prison Mine Hold 3 Jailers Hostage," *Washington Post and Times Herald,* July 14, 1959, 3.

49 "Booby Traps Discovered in Prison Mine," *Chicago Daily Tribune,* July 16, 1959, 4.

50 "Three Foremen Held Hostage in Prison Mine," *Washington Post,* November 26, 1964, A3.

51 Kushner, "Cormac McCarthy's Apocalypse."

52 June 22, 1981. Woolmer Collection, Box 1, Folder 4. Southwestern Writers Collection, The Wittliff Collections, Alkek Library, Texas State University-San Marcos.

53 "Gunman Had No Apparent Motive in Killing 3 on Gay St., Police Say," Knoxville News-Sentinel, May 12, 1976, 1.

54 "Shot Deputy Improves," *Knoxville News-Sentinel,* June 16, 1981, 1.

55 David Sowell, State of Tennessee, Department of State, Tennessee State Library and Archives, "Prison Records, State of Tennessee, 1831–1992 [Finding Aid]," Record Group 25, https://sos-tn-gov-files.tnsosfiles.com/forms/TENNESSEE _STATE_PRISON_RECORDS_1831-1992.pdf, n.p.

56 Cf., Cole Sullivan and Grace King, "'He Was Too Young to Die'— Records Show 60+ People Have Died in East Tennessee Jails Since 2016," *WBIR/10News,* November 4, 2021, https://www.wbir.com/article/news/ investigations/more-than-60-people-have-died-in-east-tennessee-jails-since -2016/51-f6d1f487-1121-4741-be5a-41aeec526d21.

57 Tennessee Department of Children's Services, "Tennessee's Juvenile Justice History," https://www.tn.gov/content/dam/tn/dcs/documents/juvenile-justice/ Juvenile_Justice_Timeline.pdf.

58 U.S. Public Health Service, *Venereal Disease Clinics, 1944 Directory* (U.S. Government Printing Office, 1944), 53.

59 Sowell, "Prison Records."

60 Tennessee Department of Children's Services.

61 "Between January and early September of 2014, there were 145 reports of violence, including 39 teen-on-teen assaults and 51 assaults by teens on staff." Jordan Buie, "Inside Woodland Hills: DCS Seeks Improvements," *Tennessean,* March 20, 2015, https://www.tennessean.com/story/news/crime/2015/03/20 /woodland-hills-improvents-sought-tennessee-department-childrens-services /25118045/.

62 In July 2019 Tennessee enacted its Juvenile Justice Reform Act, which requires, among other things, "provider performance-based metrics." The act still allows a broad reading of "seclusion," and according to some media reports,

chaos still reigns across the system. Ben Hall, "Broken: Juvenile Detention Center Locks Children in Solitary Confinement," *NewsChannel 5* Nashville Investigation, November 19, 2019, https://www.newschannel5.com/news/newschannel-5-investigates/broken-juvenile-detention-center-locks-children-in-solitary-confinement.

63 Transcriptions throughout by the author.

64 Norris Lake was completed in 1936 as part of a TVA impoundment and flood control project. The Tennessee River–spanning Watts Bar Dam, started in 1939, was completed in 1942, and served the needs of the electrical power-hungry Oak Ridge National Laboratories during the Manhattan project.

65 I am grateful to Grayson Ressler for reminding of me this passage in Wesley G. Morgan's "The Route and Roots of *The Road*," a paper Morgan presented at a seminar titled *The Road Home: Cormac McCarthy's Imaginative Return to the South* on April 26, 2007. https://trace.tennessee.edu/cgi/viewcontent.cgi?article=1002&context=utk_mccarthy.

66 Emyr Estyn Evans, *Irish Folk Ways* (London: Routledge & Kegan Paul, 1957), 59.

67 Ibid., 71.

68 Timothy Andersen, "Quantum Wittgenstein," *Aeon*, May 12, 2002, https://aeon.co/essays/how-wittgenstein-might-solve-both-philosophy-and-quantum-physics.

69 Woodward, "Cormac McCarthy's Venomous Fiction."

5

Math

Unified Theories and Fractured Minds—Toward No Probable Conclusions

Kuan, we are told, conveys a double meaning: "to be seen" or "to be contemplative." This duality, I suspect, implies an enigma of the observer or   the observed. Do we know whether what we see is a   world? Or is it a world that mirrors whatever we are?   A physicist asks if we are all alien holograms. Or, I ask,   thoughts in the mind of God?

—DONALD BEAGLE[1]

In the end, she had said, there will be nothing that cannot be simulated. And this will be the final abridgment of privilege. This is the world to come. Not some other. The only alternate is the surprise in those antic shapes burned into the concrete.

—*THE PASSENGER* (382)

This chapter follows an unconventional structure, partly the result of an effort to accomodate McCarthy's newly available (at the time of submission) 2022 duology (*The Passenger* and *Stella Maris*).[2] It begins with a consideration of the intellectual kinship of Cormac McCarthy and Guy Davenport, including their far-ranging correspondence, two unified minds questing for a unified aesthetic theory. Then it turns to the "timeless" language of mathematics and the duology novels, which McCarthy has described as a study in madness and genius. Since the two new novels' explication could fill volumes, this section offers a brief overview with examples of how to read them as a disciplinary code-switcher, as well as the clues that McCarthy offers for how to read them. The chapter ends with a coda, a thought experiment or (at risk of seeming twee) an experiment for unthinking and unknowing, really, that hits at the late interest in simulation theory, McCarthy's lately found interest in it, and some thoughts on how, in the unmaking of worlds, we find their expansion.

*   *   *

In *Stella Maris* (2022), the protagonist/mathematician, Alicia, has a bone to pick with our quotidian spoken language as distinct from mathematics. In her view, the spoken word is a sort of dissembling, come-lately substitute: "In the end this strange new code must have replaced at least part of the world with what can be said about it. Reality with opinion. Narrative with commentary" (175). I do not propose to offer here a poetics (a study of linguistic technique) binding narrative and science, or mathematics and narrative, for reasons explained earlier, not least of which is that McCarthy's duology, *The Passenger* and *Stella Maris,* can be read expressly as just such a poetics, as Alicia's musings show. Instead, I offer a compromise, a shared quest that emerges from the correspondence of Cormac McCarthy and his longtime friend Guy Davenport (1927–2005), exploring the possibility of an aesthetic theory of everything—and perhaps of nothing in the sense of *ex nihil* originalism—in McCarthy's expanding worlds.

To the extent that the meeting of their minds advanced these ideas, McCarthy must share his prize for co-discovery with Davenport. An intellectual jack-of-all-trades, Davenport was an internationally respected man of letters. The son of Guy Mattison Davenport and Marie Fant, he was born in Anderson, SC, on November 23, 1927, and showed early literary inclinations. At twelve years of age, he created his own "newspaper," *The Franklin Street News*, precociously reporting on "visits, birthdays, the births of kittens and puppies." As a child, Davenport received private tutoring at Anderson College. He went on to study at Duke (BA, 1948) and Oxford (BLitt, 1950) on a Rhodes scholarship, where he wrote his thesis on James Joyce's *Ulysses*. This fact alone might have endeared him to McCarthy, a fellow Joyce aficionado, but Davenport, like James Dickey, had additional charms, dimensions of character formed by military service and wide travel. He served with the Army Airborne Corps (1950–2) and began his teaching career at Washington University in St. Louis (1952–5). In 1952, while conducting research for an article, he formed a lasting acquaintance with Modernist poet Ezra Pound, subsequently visiting the writer during his imprisonment at St. Elizabeth's Hospital for the Insane. When Davenport completed his PhD at Harvard (1961), Pound was the subject of his dissertation, and a frequent subject in his correspondence with Cormac McCarthy, which started in the 1960s.

Davenport considered himself a teacher foremost—McCarthy's early letters address him as "Dr. Davenport"—and his writings as "an extension of the classroom," representing the creative component of a searching mind. He started teaching at the University of Kentucky in 1963 and retired from there not long after receiving a MacArthur Foundation Grant in 1990. Davenport wrote prolifically in multiple genres and served as an illustrator for his own books and others. He authored more than twenty of them, including two poetry collections, while publishing book-length translations of classical texts by Sappho, Diogenes, and Heraclitus, the last being a notably large influence on McCarthy's work and philosophy. Davenport's articles appeared in *Virginia Quarterly Review*, *Life*, *New York Times Book Review*, and *National Review*, where he was contributing editor (1962–83). His interest in the history of ideas and his broad learning imbued his fiction with its rarefied texture: it was densely allusive, frequently experimental, and classically informed, a Modernist criticism tooled to Modernist writers. Davenport described his stories as "lessons in history," requiring close reading, appropriate for a man who proclaimed that "[a]rt is always the replacing of indifference by attention"—a virtual restatement of McCarthy's credo on curiosity's value (see Chapter 1).[3]

Davenport was unassuming by nature, and when John J. Sullivan interviewed him for the *Paris Review*, Sullivan responded incredulously to Davenport's claim that "I have no life" by exclaiming, "This from no doubt the only man on earth to have known Ezra Pound, Thomas Merton, and Cormac McCarthy personally."[4] The correspondence between McCarthy and Davenport sheds light on a warm friendship grounded in true familiarity, mutual peer review between writers, and in-person visits paid through the years.[5] Some of McCarthy's letters are blunt, like a 1979 missive expressing his distaste for modern writing in general that foreshadows later musings on linguistics: "And [modern poetry] dont sing. Surely poetry is a bastard offshoot of song anyway. In the larger scheme of human undertaking I suspect it has been around but a short while and may not be around too much longer."

The letter is of particular interest here because it shows that McCarthy was already reading in physics—a first step along the path to SFI—in the 1970s, and already on his way, ironically, to holding language in suspicion.

It offers a rare glimpse into McCarthy's interest in translational science, as well as an aesthetic manifesto that carves out a separate standing for math and science. The letter continues,

> Song and story seem to be indigenous to the soul of man, but I wonder what it means that literature has become so estranged from people's lives. Rutherford said that modern physics should be capable of being put into language that a scrublady could understand. Well, maybe. I think I'd rather undertake explaining, say, Heisenberg's theory of indeterminacy, than Ezra's poetry. But that's not really the point. Physics is concerned with framing a description of reality and it can maintain a tradition, however esoteric, as long as there are people interested in it. Literature, on the other hand, lacks this purity altogether, and as the rise of the masses make itself felt on an ever increasing scale, those who pursue, or affect to pursue, excellence in letters are driven to higher ground where they scrabble about on the bones of dead and cultish vogues that no longer have any rooting in the lives of people.[6]

The letters are revealing in other ways, too. McCarthy enthused over *Surely You're Joking, Mr. Feynman!* (1985)—"The most fun I've had reading in a year"—long before he would help edit Lawrence M. Krauss's 2012 biography, *Quantum Man: Richard Feynman's Life in Science*, at SFI.[7] In one letter McCarthy reported that he met Davenport's "friend Steven [*sic*] Gould." He meant Stephen Jay Gould (1941–2002), the well-known evolutionary biology and popular science writer, whom he describes as "an interesting man but a thoroughgoing materialist. An explainer, in short."[8] Other letters reveal that Cormac spoke to Ralph Ellison on the phone, was steered by Henry Miller to Charles Montagu Doughty's *Travels in Arabia Deserta* (1888) and Oswald Spengler's *The Decline of the West* (1922), and wondered aloud if Davenport "had a hand" in the MacArthur Fellowship he received in 1981.[9] When Davenport expressed disappointment in Red Ozier Press and lukewarm reviews of his essays, McCarthy consoled him, "[T]hose essays are as good as any written in this country to date." In a 1986 letter he deems Davenport "the best living practitioner" of the literary essay.[10]

At the height of their correspondence, Davenport and McCarthy were both on their way to producing not just a mutual admiration society but respective masterworks—McCarthy's *Blood Meridian* (1985), and Davenport's *Every Force Evolves a Form* (1987). The latter collects sixty of Davenport's highly regarded critical essays, touching on such thinkers as Ludwig Wittgenstein (again, a large influence on McCarthy), Charles Olson, and Samuel Beckett.[11] McCarthy's attitude toward Davenport's critical essays, at least insofar as it emerges from the Ransom Center letters, verges on the uncharacteristically worshipful. Over the span of several decades, McCarthy pestered his local bookstores to stock Davenport's esoteric work "in hardback and paperback," and made sure they carried "multiple copies," as their correspondence documents.[12] McCarthy tried also to expand Davenport's sphere of influence and to connect him with large publishing houses. His estimation of Davenport's best-known work of fiction, a wildly creative and atemporal novella called *Tatlin!*, was typically restrained—he was most comfortable providing a line edit—and yet encouraging.[13]

In the 1980s Davenport was already writing about what we now call text-mining and textual visualization as a pathway to discerning textual unities, feeling his way to a humanistic theory of critical complexity. In "The Critic as Artist" (1985), he surmised, "This conjunction of writer and critic is apt to surprise us in larger measure than we can anticipate":

> Recently Ralf Norrman and Jon Haarberg, two Scandinavian critics handy with structuralist strategies, asked themselves what would emerge (by way of crystalline geometric figure floating above the spatter) if they made a survey of pumpkins, cucumbers, squash, and gourds in world literature. The result was happy, fascinating, and wonderfully surprising. They were able to write something like a formula for the iconographer to follow: that where these vegetables appear in imagery, certain meanings come with them, and certain tones are achieved. They go so far as to ask, quite seriously, if literature writes itself. Their theory approaches what we have all suspected, that culture is a language of images and ideas, and functions according to a syntax, with dialects and idiomatic constructions. . . . When the critic can specify two spatter patterns, and hand us the stereopticon, to make the magic figure appear in the middle of the air, he is indeed entering the artistic process.[14]

Or in other words, he is on his way to complex systems science applied to literature, and from thence to GPT-3 (Generative Pre-trained Transformer3), an AI that uses deep learning and a language model to write remarkably convincing prose, bringing us to a place where literature can indeed write itself.[15] Norrmann and Harrberg's methodology has been supplanted by readily available text-mining digital humanities tools, including the Google Ngram Viewer—for those who care to look into it. Yet Davenport was truly ahead of his time inasmuch as his essay offers multiple examples of complex, deliberate density in literature:

> From *The Cantos*, William Carlos Williams' *Paterson*, where the difficulty is in subtlety of technique rather than in subject matter. From *The Cantos*, Louis Zukofsky's "*A*," where the difficulty of reading is immense, and a new kind of difficulty in any kind of poetry. From Pound's *Cantos*, the *Maximus* poems of Charles Olson.

All of these texts are difficult. The stock explanation for this is that modern life is complex: an art reflecting it must be equally complex. This insight is a bit blind, it seems to me.

It would make more sense for a complexity of life to call into being an art of great lucidity, one capable of countering confusion with clarity, one that might ascertain certainties in a chaos of uncertainty, to comfort and guide us. Indeed, that is precisely what Pound said he was doing in *The Cantos*, "cutting through the muck with clarity." If we look at the programs of the most difficult of modern writers, we will find them to be unaware, to a surprising degree, of the problems created by their work.[16]

McCarthy praised to the stars Davenport's earlier critical essay, "The Geography of the Imagination" (eventually the title essay of his 1981 collection), both in correspondence with the author (he asked Davenport to sign his copy), and separately, to Howard Woolmer, as a first-rate work of literary criticism.[17] In that essay, Davenport advanced a theory in which entropy in art increases over time, and, accordingly, suggested that the critical enterprise should be one of divergent diminishing complexity, in which unities can be discerned. (As with complexity theory, Davenport's theory might be said to rely on a grammar which is evolutionary, whereas language is not, if McCarthy's assertion in "The Kekulé Problem" holds true.)

McCarthy and Davenport were both keenly interested in whether language was underpinned by unchanging, non-evolutionary realities, an *ur*-language that could be divined by critic and artist alike. Some of those unities might still operate in language through archaic survivals whose meaning has changed so much as to be concealed from its own speakers— or at least, some linguists at SFI have suggested so. *The Counselor* (2013) can be read with utter consistency as an ironic play of truths spoken in jest or simply inadvertently, sometimes in a version of the Freudian slip, revealing the contemporary confusion of the sacred and the profane, and the unconscious manifesting itself awkwardly through language, using even malapropisms to surface truths. By the time of the duology, McCarthy had honed these illustrations into the hard example of the Thalidomide Kid.

Accordingly, the Santa Fe Institute continues to designate the Evolution of Human Language Project as a project—not a theme per se—tracing its origin to a conference on Linguistic Databases and Linguistic Taxonomy (January 6–10, 2003) organized by Murray Gell-Mann and Sergei A. Starostin, with continuing work most recently spearheaded by Tanmoy Bhattacharya.[18] Part of the debate that continues to swirl around the topic is epistemological: could such a thing ever be retrieved or definitively knowable? A majority of linguists remains skeptical that scientific method could be applied to comparative linguistics this way, since etymologies are themselves a series of unattested opinions as part of an ongoing battle of the experts who tend to be intrinsically biased by their mother tongue. Whether this makes the SFI investigation peripheral, necessary, or simply starry-eyed depends on which expert one consults. Noam Chomsky believes that there could be a single theory of language, yet as a proponent of discontinuity he asserts that human language commonalities owe much to how our brains are hard-wired to receive language, which in turn stems from a single random genetic mutation for language facility in the brain that arose *c.* 100,000 years ago, passed along rather splendidly (as distinct from the long-term cultural evolution of language). Of course, all linguists who investigate the origins of language face a Russell's teapot problem: Is this a falsifiable claim, and who can disprove it?[19]

In any case, the 2003 Santa Fe gathering drew on the work of linguists who had tried to make a case for linguistic monogenesis—a single common ancestral tongue—including scholars like Meritt Ruhlen and John Bengtson, who, for example, believed that they could point to words with common roots across all language phylums. Another fruit of the 2003 gathering was a paper published by Gell-Mann and Merritt Ruhlen in the *Proceedings of the National Academy of Science* that postulated that the proto-human-language followed subject–object–verb (SOV) word order (think of Star Wars' Yoda syntax).[20] However, about half of the world's languages retain this order, or capriciously cycle it, so where is the expected evolution from the original? And can the question ever be answered definitively? In many respects, the line of inquiry seeks the branches in the evolution of human languages, and a common ancestor, a quest in paleontology that is clouded by the readable life of DNA in the fossil record, and the complex blending and extinction of certain lines. Nevertheless, for now SFI still nourishes the Evolution of Human Languages program "and its associated data banks in historical linguistics," preserving a line of inquiry into "the origins, evolution, and diversity of human languages."[21] SFI's inquiry in this area has taken a computational turn toward evolutionary biology, with efforts to use artificial intelligence to reconstruct the ancient linguistic past:

> We believe that evolutionary processes do give rise to cultural diversity, and we seek to uncover and describe these processes through logical and mathematical analysis of human language. The goal of our project is to go beyond the merely abstract similarity between biological evolution and cultural evolution to develop a quantitative framework for understanding the relatedness of human languages that incorporates the vast experience accumulated over two hundred years of research.

AI and gene mapping have made it possible to authoritatively sort out genealogical trees in ways that look beneath beguiling phenotypical similarities and beyond false taxa (consider how new molecular technology eventually vindicated Vladimir Nabokov's first-doubted 1945 hypothesis that a group of South American butterflies could be traced to Southeast Asian ancestors).[22] With enough linguistic inventory the same taxonomic methodologies might be applied to divine the deepest linguistic relationships, permitting complexity scientists to peer beyond polysemous and superficially related "organisms"—or so it is hoped.

In "The Kekulé Problem" (2017), McCarthy wades into this linguistic debate with a remarkable essay. SFI president David Krakauer's foreword to the article explains that McCarthy is

> [a]n aficionado on subjects ranging from the history of mathematics, philosophical arguments relating to the status of quantum mechanics as a causal theory, comparative evidence bearing on non-human intelligence, and the nature of the conscious and unconscious mind. At SFI we have been searching for the expression of these scientific interests in his novels and we maintain a furtive tally of their covert manifestations and demonstrations in his prose.
>
> Over the last two decades Cormac and I have been discussing the puzzles and paradoxes of the unconscious mind. Foremost among them, the fact that the very recent and "uniquely" human capability of near infinite expressive power arising through a combinatorial grammar is built on the foundations of a far more ancient animal brain. How have these two evolutionary systems become reconciled? Cormac expresses this tension as the deep suspicion, perhaps even contempt, that the primeval unconscious feels toward the upstart, conscious language.[23]

I would submit that the unconscious draws us closer to the unmentionable, the deepest matter in our evolutionary firmament—creation, destruction, and our small sentience within a universe to which we might ascribe sentience. For all its creative power, language, as McCarthy suggests in *Whales and Men*, also expelled us from the garden. Recursive and sometimes autophagous, the evolutionary loop, in which ruder forms, remarkably, survive, is also the loop of language, a Hegelian process. Most of us go through life trying to sound out the universe for what is true because we are listening for the whisper of what is good. We are all subject to the confusions of anger and desire, and yet we remain attuned somehow to the insistent lessons of love. In the post-apocalyptic world of *The Road*, the father who would lead his son to safe harbor thinks, "On this road there are no godspoke men. They are gone and I am left and they have taken with them the world" (27). *Godspoke* is a word of McCarthy's own vintage, and a wonderfully useful one, since words call things into being, and lose their meaning in a world unmade. The godspoke—those brought into being by the *logos*—are the creative force in the world. They stand in contrast with the rickety cannibal encountered by the novel's father and son, a destroyer "who has made of the world a lie every word." Truth has the opposite effect, encouraging us to call the good into existence, even when, at times, we cannot yet see it, or hear it above the noise, and McCarthy points to its grounding in a wordless unconscious and the metaphysically ineffable (cf., John Grady Cole in *All the Pretty Horses*: "There aint but one truth. . . . The truth is what happened. It aint what come out of somebody's mouth" [168]).

What emerges from the words between Davenport and McCarthy, and the sparking of their minds, is a way to think about looking for unities in complexity, of taking the joyful disciplinary leap, and understanding literature from within. Writing about his friend and correspondent, the poet Louis Zukofsky (1904–78), Davenport wonders,

> Was Zukofsky in this last book teaching us something he felt we needed to know? *Finnegans Wake* is full of instructions as to how it is to be read, and the allusions in *The Cantos* turn out to be something he felt reinforce each other, so that the more we know about them, the more we see how they fit together. It is precisely this sense of field, or family, that makes modern literature different. And it is the business of the critic to be able to say where the boundaries lie, and where the center is. For the tacit agreement between writer and reader maintains—it is the order by which we can read at all—but it has undergone in our century a change whereby writing has had to insist over and over that it is, as always, words in a pattern. This pattern has always fitted, in hundreds of different ways, the pattern of the world.[24]

I agree, and more, I think we can apply this insight to McCarthy's duology, the two new mathematical/scientific novels that, like *Finnegan's Wake*, are full of instructions on how they are to be read, and that promise to be a source of both fascination and frustration for readers well into the future.

*   *   *

A draft had been glimpsed after it was inadvertently placed among McCarthy's open archival papers. Rumors of the imminent publication of a single extended work—which turned out to be two—had circulated for over a decade. In August 2015 the Santa Fe Institute dramatized portions of it, replete with Cormac's voice-overs, an event that was noticed around the world,[25] and with the seal broken, it was widely assumed that the work would soon be published.

And then—nothing happened.

The year 2020 tolled in a new decade for a manuscript that some said McCarthy had been working on ever since arriving at SFI, twenty years earlier. Had the clacking of his Lettera 32 Olivetti typewriter, frequently heard in the Institute library before it was auctioned for $254,000 in an

SFI fundraiser, simply been McCarthy typing his version of Joe Gould's secret? McCarthy's own family members could not be sure if or when the book(s) would be published (I know this because I talked to several of them about it).

And then . . . something finally happened.

In March 2022, the *New York Times* and other outlets announced that Knopf would publish the two novels later in the year.[26] The *Times* teaser might lead readers to think that McCarthy had penned another *No Country for Old Men*-style potboiler with *The Passenger*:

> *The plot is set in motion when Bobby, a salvage diver, gets assigned to explore the wreckage of a sunken jet off the coast of Mississippi, and discovers that the plane's black box, the pilot's flight bag and the body of one of the passengers are all missing. With the pace and twists of a thriller, the 400-page narrative follows Bobby, who is haunted by his memories of his father and sister, as he gets drawn into the mystery of the plane crash, and realizes he may have uncovered something nefarious when strange men in suits show up at his home.*

And then . . . in my summary . . .

Nothing happens. The potboiler stops boiling.

The précis covers around the first 80 pages or so of the novel, which becomes a study in Bobby Western's solipsistic wanderings into existentialism and concurrently his weakening grasp on reality. In an enthusiastic early review, Jonathan Miles writes, "And though a pair of federal agents comes sniffing around, and a few more riddles bubble forth, the sunken jet essentially serves as a four-engine MacGuffin that fades out by midnovel."[27] Like a discarded Nike listlessly floating its way to the Great Pacific Garbage Patch, the novel is borne with sometimes unbearable slowness on currents we cannot easily see. This is a pejorative simile, of course, but it takes a lemon chaser to swallow the cheap whiskey of *Times* marketing copy, because (contra the puffery) the next 300 pages have nothing of the pace or twists of a thriller. McCarthy demurred to describe *The Passenger*, which was still a one-volume work in progress, in his 2009 *Wall Street Journal* interview ("I'm not very good at talking about this stuff"), but in fact, it is difficult to improve upon the concision of the summary he offered then: "It's mostly set in New Orleans around 1980. It has to do with a brother and sister. When the book opens she's already committed suicide, and it's about how he deals with it. She's an interesting girl."[28] The first three sentences cover the last 300 pages of *The Passenger* in which very little happens, and the last sentence summarizes the 180 pages of *Stella Maris*.

To top it off, the Thalidomide Kid, a malformed figment of apparently hallucinatory "imagination," strides into *The Passenger* on the fourth page, "*kneading his hands before him like the villain in a silent film. Except of course they werent really hands. Just flippers. Sort of like a seal has*" (5). The Kid will reappear at intervals throughout *The Passenger* and with even greater frequency in *Stella Maris*, playing the part of Vaudevillian interlocutor, spouter of malapropisms, and minstrel fool. I anticipate that most readers will find these italicized sections, voiced by brain goblins, at best, trying, and besides offending certain contemporary sensibilities, they are likely to be called out for lacking verisimilitude by people who study and who have experienced mental illness/schizophrenia. It's a damned difficult point of view to write, even for the people who have been there, apparently. Fundamentally (if not satisfactorily for scientific materialists) one way to understand schizophrenia is that, with so many circuits open, the unconscious begins to hijack the conscious mind (see, for example, Barbara O'Brien's first-person account in *Operators and Things: The Inner Life of a Schizophrenic* [1958]).

McCarthy has a built-in defense to these charges, because we are meant not to know whether the Kid is "real," and most readers will misidentify him, allegorically speaking, which is the very point McCarthy wishes to make—when the unconscious speaks to us, we are likely to perceive it as grotesquerie, instead of examining the dumb show of our lives, dreams, and most fugitive thoughts. McCarthy indeed has already taught us how to read *The Passenger* and *Stella Maris*, which are very much novels of ideas—even more so than *The Sunset Limited*, or the screenplay *Whales and Men*, rendered commercially unfilmable for want of dramatic suspense, or *The Counselor*, widely regarded by critics as a problematic screenplay that led to a meretricious film (though I think the work is a metaphysical serio-spoof that flew over the heads of reviewers and audiences alike). The Kid offers an extended illustration of McCarthy's reasoning in the Kekulé essays. By my theory, the Thalidomide Kid is primarily an elaborate device, albeit a sometimes exasperating one: the unconscious grabbing the conscious by the collar, to which we are mostly resistant (and this includes sibling Bobby Western, the protagonist of *The Passenger*).

And now, reader, comes the moment for a pop quiz, or at least a flashback: why is thalidomide an important compound in the chemical understanding of chirality (hint: reread Chapter 2)? Because the chiral twinning of the same molecule can have very different effects in its interactions. Flip it one way, and you have a mild sedative; flip it the other, and you have a compound that causes severe birth defects from scrambling the instructions for the programming of life.[29] Flip it one way and you have consciousness, and the other, the unconscious, whose appearance is distorted, warped, and often shunted aside because of language's arrogation of a neural network that did quite well without it for most of human existence. Flip it one way (perhaps by negotiating and reconciling the unconscious and the conscious), and you have genius. Flip it the other way, and you have madness, the place of most profound understanding locked behind hopelessly communication-negating walls, the anguish of Alicia's imprisonment by understanding, replacing "Reality with opinion. Narrative with commentary" (175).

Thalidomide is also the failure of science to understand the human adequately, and the human failure to understand science adequately. Notice that the Kid rides again with *The Passenger*; the protagonist of *Blood Meridian*, the lower-case-k kid, grapples on an ahistorical plane with the destructive element in the universe, whereas the (Thalidomide) Kid attempts to protect Alicia from her own self-destructive impulses. McCarthy is a creator who tends to use the pieces on the floor, and it might be pointed out that he reprises symbols and themes across works in a distinctive way that brings a truly pleasing wholeness to his canon. For example, in *The Crossing* he raises a set of metaphysical questions around an indigenous salvage crew, led by a man referred to simply as "the gypsy," who have been commissioned by a bereaved father to retrieve the remote, Chihuahuan desert-bound wreckage of a cloth biplane, a memento mori of the American's son who died in the craft. The efforts beget additional misery, ruination, and loss of life among those referred to as *men of the road/path*. "In those first years," McCarthy writes elsewhere in *The Road*, "the roads were peopled with refugees shrouded up in their clothing. Wearing masks and goggles, sitting in their rags by the side of the road like ruined aviators" (28). We return to the men of the road, and the figure of the aviator. In the world of *The Passenger*, we have another plane down, another ruined (or missing) aviator. The Icarus symbol seems to have a death-grip on McCarthy's imagination, perhaps as a portal to the unconscious and a set of attendant metaphysical questions.

Similarly, the evasive manifestations of the Thalidomide Kid flicker across works. In my reading, at least, it is to apply the Kekulé framework to understand this bizarre and insistent homunculus. The problem, of course, is that very few will have the patience for all that, or *Stella Maris*'s rolling scrolls of historic mathematical arcana. The Joycean verbal gymnastics and punnery are good fun but likely not enough to sustain the act except for dedicated logophiles and quantum nerds (a class to which I cheerfully consign myself). In keeping with the Kekulé framework, the Kid energetically furnishes malapropisms and Freudian slips, cursing/swearing and adjuring and the like, giving flashes of how the unconscious runs interference even in waking hours to fluster the censorious man upstairs. The Kekulé essays suggest, darkly, that the unconscious might not be good, and could even be malign for all we know, consistent with McCarthy's gnostic notion that our existence might take place in a warped, evil counterfeit of a better world. Alicia Western, the protagonist of *Stella Maris*, tormented to her eventual suicide by incestual desire and the profound alienation of her genius, cannot be saved by the Thalidomide Kid, the hallucination that remains her constant frenemy as she checks in and out of mental hospitals. Yet it should be noted that the Kid is invested in saving her from toothy perverted men and other real threats, including her self-destructive impulses. Like Black in *Sunset Limited*, he simply fails in the end. The Kid is a machine for operating an animal that has become entirely cerebral, unable to reconcile narrative with deeper realities glimpsed in mathematica.

In a troubling respect, *The Passenger* at times seems foundered by the problem of having too much information, and the author seems precariously close to slipping into the madness of his characters. If we can even try to construe a plotline, it slips away by the end: the significance of the missing passenger, which sets up the mystery of the text, emerges neither in *The Passenger* or nor in *Stella Maris*. Instead, a series of apparently unconnected incidents which only paranoia (or quantum entanglement) might connect: a very unsettling stay on an oil rig off Pensacola goes unexplained, and mysterious characters come and go, including an intercessor named Kline, for purposes we cannot discern, reaching a low point when Bobby rehashes conspiratorial Kennedy assassination theories in the novel's later pages, its speculations descending, like a coastal shelf, to strata upon strata of dirty money.

Of course, as the saying goes, just because you're paranoid doesn't mean that everyone isn't out to get you. Pointing to the significant part of American economic output that has always been entangled in dirty money, economist Paul Krugman wrote in the *New York Times*, "There's about $1.6 trillion worth of $100 bills in circulation—80 percent of all U.S. currency—even though large-denomination bills are very hard for ordinary consumers to spend. What do you think people are doing with all those Benjamins?"[30] About what one might expect: pumping them through the black markets that comprise a significant part of the American GDP, that created south Florida in the twentieth century, and that gave Carlos Marcello, a person of interest in *The Passenger*, a $2 billion stake in his mobster reign over the Crescent City for a time. But where McCarthy wants to take these ideas remains inscrutable at the novel's end, which finds Bobby climbing into his version of the Martello tower in Ibiza, having earlier endured exile in frozen Idaho and a desolate shack on the Gulf Bay. Places of perhaps real and numinous importance in McCarthy's life, yet of untranslatable and doleful significance here, or else, to borrow Hemingway's signal-phrase for depression, excursions into the black-assed swamp of the past.

An appropriate title for the first volume would be "Fragments of the 1980s," given the clearly autobiographical proximity of Bobby Western to Cormac McCarthy, not to mention the many appearances of the identifiable living and the dead (including McCarthy's off-beam, homegrown parlor intellectual from Knoxville, John Sheddan, an irredeemable scammer, bounder, rounder, and persistent friend). So we have fragments of other texts, of people and places visited, of relationships reworked, yet that could not be reworked into sense-making narrative. It is easy to find fault with *The Passenger*, yet early advance reviewers have seemingly been reluctant to admit bafflement, and have been mostly positive, citing notably vague virtues in the text. As McCarthy said, self-deprecatingly, on NPR, "I'm pessimistic about a lot of things, but as Lawrence [Krauss] has quoted me as saying, there's no reason to be miserable about it." He goes on to say, "The other thing we talked about a few minutes ago was how bad we are at prognostications. So the fact that I take a pretty dreary view of the future is cheering because I think, you know, the chances are that I'm wrong."[31]

Along those lines, I am not miserable about the two novels, and bow to my many respected colleagues who are charmed by them. Though the books might not offer the most compelling swansong, they will do for a *Finnegan's Wake* to keep the critics busy for a hundred years, and they have a place of integrity within McCarthy's larger opus. Anticipating some common themes of the critiques to come: the antagonistic voices that create dialogue—contrary personalities, Black/White, Alicia/the shrink, the Kid/everyone else—seems to be a mainstay of Cormac's way of questioning reality. Here they come to sound largely the same and often pointlessly antagonistic. Their inability to pull back the metaphysical curtain grows tiresome if readers cannot come to care about them and their struggles. Consequently, the characters and their philosophical dialogues are poorly differentiated. The voices run together, and so do the conversational loops, tendentiously toward an expertise that seems to belong more to the author than his characters.

Yet *The Passenger* still has something to offer, too, especially as ancillary to *Stella Maris*, the dialogic novel I will turn to now, with the aim of showing how it might be read, or how McCarthy has signified that it is to be read. The novel is entirely a dialogue between Alicia Western and her psychological examiner. It is, as Alexandra Alter observes, "the first time that McCarthy has built a narrative around a female protagonist."[32]

This is true, and more, Alicia Western has a sharp awareness of her place in mathematics as a gendered field. Dr. Cohen asks, "Are most of your heroes mathematicians?" and she replies,

> Yes. Or heroines.
> Who else do you admire?
> It's a long list.
> Okay.
> Cantor, Gauss, Riemann, Euler. Hilbert. Poincaré. Noether. Hypatia. Klein, Minkowski, Turing, von Neumann. Hardly even a partial list. Cauchy, Lie, Dedekind, Brouwer. Boole. Peano. Church is still alive.

Hamilton, Laplace, Lagrange. The ancients of course. You look at these names and the work they represent and you realize that the annals of latterday literature and philosophy by comparison are barren beyond description.

Those names are not familiar to me.

I know.

Are any of them women?

Emmy Noether. She was a great mathematician. One of the greatest. One of the founders of mathematical physics. There are others. Women. No Fields Medals yet of course. (67)

Elsewhere, Alicia recounts her time as the first woman to be invited to the Institut des Hautes Études Scientifiques ("I was the only woman there. At first they thought I worked in the kitchen" [11]). McCarthy clearly has some awareness of the delicacy of what he is essaying as a male author. In the *Wall Street Journal* interview, when told, "Some critics focus on how rarely you go deep with female characters," he responded, in part, "I was planning on writing about a woman for 50 years. I will never be competent enough to do so, but at some point you have to try."[33] Why he must try is not explained. Alexandra Alter contends that in *Stella Maris* "he inhabits the shattered psyche of Alicia Western, a math prodigy whose intellect frightens people and whose hallucinations appear as characters, with their own distinct voices."[34] The extent to which McCarthy "inhabits" a female consciousness will no doubt be another topic of much discussion, and to say that Alicia has hallucinations, as Alter contends, is problematic, too, as this exchange shows, beginning with Alicia:

That a drug can restructure the world into something like an objective reality is a claim with as little validity as the objective reality itself. I think what I said at the time was that I had no more reason to place my confidence in a drugged state of mind than in a sober one.

You wouldnt be willing to try another medication.

You asked me that.

All right. If someone were to come into the room while the Kid was there might they see him?

And that. But probably not.

But not exclusively.

I dont know.

If they were on the reality drug with the rest of us then I suppose not. (104)

If Alicia's reality as respects social norms is so eroded that she cannot conform with so-called natural law and fails to see why an incestuous relationship with her brother is proscribed, it is difficult to know what to make of her—but, then, it is also difficult to know where the rule comes from if it is not biologically encoded in the form of instinctive revulsion. "What it all adds up to—perhaps surprisingly," writes Jonathan Miles—"is a doomed and unsettling love story, a Platonic tragedy." Alicia and Bobby Western are the only two people on earth capable of understanding each other and grokking the world of platonic forms—tough luck, that. In conversation, Dennis McCarthy has affirmed this way of reading the novel to me, seeing its greatest power as a moving love story in a tragic register, a pathos that I admit was wasted on me. Although this is not the first instance where Cormac McCarthy has brought incest into the inner, visible dark of his fictional universe, my first instinct was to find a way out for Alicia. I strained for several possible readings, based on textual clues, that might at least open the possibility that she is Bobby's half-sister, or purely adopted sister. I cast around for candidates for real-life models, including Katherine "Toni" Oppenheimer, Robert Oppenheimer's daughter who killed herself in her early thirties.[35]

But even if this riddle can be eventually solved definitively, it will not teach us how to read *Stella Maris*, or resolve the moral question at the center, and it might only prejudice the reading. So instead, we might turn to American philosopher Willard Van Orman Quine, who Alicia says might be the greatest philosopher of our time, and who, slyly and only half-facetiously, used the Sherwin Williams paint slogan—*Save the surface and you save all*—as the epigraph to one of his best-known works. Here is Quine on the nature of reality and what is:

> A curious thing about the ontological problem is its simplicity. It can be put in three Anglo-Saxon monosyllables: "What is there?" It can be answered, moreover, in a word—"Everything"—and everyone will accept this answer as true. However, this is merely to say that there is what there is. There remains room for disagreement over cases, and so the issue has stayed alive down the centuries.[36]

As James Brown would say, Take me to the bridge. We are at the bridge now. The dialogue of *Stella Maris* brings this "disagreement over cases" to life. Our young, distraught genius is radically skeptical of everything in the universe. The Thalidomide Kid, to my way of thinking, is the messenger from the other side of reality, and particularly metaphysical reality, which requires neither movement nor matter for understanding. How is language servant to, adequate to, or otherwise implicated in expressing that side of reality? McCarthy flirts with the idea that mathematics and perhaps physics are a better form of expression—the best of all languages for describing what exists. The mathematical Platonists believed math to be capable of metaphysical expression, which really would elevate it to the language of the universe. But *Stella Maris* goes into all that's problematic about such a notion. The unconscious, which has co-evolved with us, holds the keys, the text suggests—speaking a language even beyond math and physics, which remain self-unraveling descriptive systems.

To hear McCarthy speak about this, having quite a bit of sport at play in the fields of the scientists and creatives, one might turn to an often-overlooked National Public Radio *Science Friday* interview bringing together, imponderably, host Ira Flatow, "filmmaker Werner Herzog, novelist Cormac McCarthy, and physicist Lawrence Krauss," who joined the show "from Tempe, Arizona, where they gathered for Arizona State University's Origins Science and Culture Festival." In Flatow's summary, "Our conversation was really wide ranging. It took us from the Chauvet cave in France, the setting for Herzog's . . . film *Cave of Forgotten Dreams*, to the frightening dystopia of the novel *The Road* by Cormac McCarthy, to the outer reaches of the cosmos." In a clip from Herzog's film, a "young archaeologist" describes how after visiting the Chauvet cave and seeing the 32,000-year-old paintings of animals on the walls,

> [I]t was so powerful that every night, I was dreaming of lions. And every day was the same shock for me. It was an emotion of shock. I mean, I'm a scientist, but a human too. And after five days, I decided not to go back in the cave because I needed time just to relax and take time to absorb it.[37]

In the conversation between the guests, McCarthy demonstrates how much he knows about ancient cave paintings, leading up to a telling exchange with Werner Herzog:

> HERZOG: Yeah. When you speak about forgotten dreams, you know, there's one stunning piece unearthed, a rock pendant. The only partial human depiction, the lower part of a female body, naked, the pubic area visible, and the bison somehow embracing the female. And 32,000 years later, you have Picasso drawing paintings and doing prints of the Minotaur and the female.
> McCARTHY: You know what Picasso said when he came up out of Lascaux after the war?
> HERZOG: Yes.
> McCARTHY: He said, we've learned nothing.[38]

So here we arrive at the real aesthetic, per the novelist, hidden in the dark heart of the world, and a part of human experience that might be prelingual and heedless of language. So much for poetics. "Ideas are technologies," David Krakauer says, in the spirit of Neil Postman's *Technopoly* (1992). "They're not physical artifacts but they're cultural artifacts that pass from one person to another and allow us to more efficiently encode reality."[39] The unimprovable images in the cave directly apprehend ideas, and in turn, are technologies which may or may not encode present realities. If there is a reality, and if our codes can express it, and if we can retain sanity in a lifetime spent questing for answers. Keeping these ideas in mind prepares us to read this novel of ideas, per Quine, as an extended instance of "room for disagreement about cases."

*   *   *

So what is meant, in the end, by McCarthy's expanding worlds? If this is not self-evident by now, perhaps a final mind-bending, or exploding, exercise will get at it. By way of illustration, and to show how emergence applies to the recurrent questions of McCarthy's works, we will embark on a thought experiment that thoroughly disregards Wittgenstein's commandment that we cannot talk about things that are not in the world. It also clashes with Quine's preference for "desert places" (eat your heart out, Cormac McCarthy) as the uncluttered starting place for ontologies, since, in my view, the negative space of the universe is surely implicated in what it is and might very well involve forces that evolve forms.

Let us begin, counterintuitively, with this statement, which winds its tortured and tortuous path across all of McCarthy's works:

God does not exist.

which, paradoxically, instantly "exists" him/her/them (here *exists* is deployed as an active verb—i.e., *calls into existence*) within a potentially infinite set of things that do not exist.

Now imagine a machine learning loop in which the rule, derived from apophatic theology, is

God is not what you say he/she/they is.

What results if this artificial intelligence is sent to discover god? Perhaps the reverse-articulation of an endlessly expanding universe *ex nihil* (from nothing) as a vast set of permutations rolls forward and each overruled in turn. In fact, artificial intelligence already employs these looping exercises to simulate reality, and to approximate, in Quine's phrase, what is. The way that generative adversarial networks generate plausible images of nonexistent human faces is through a looping zero-sum game pitting a generative network against a discriminative network. It emulates the familiar *hot-or-cold* guessing game based on available data: the randomly generated face is either getting closer to a realistic human countenance, or moving away from it, based on feedback from the discriminative network (informed by a very large set of real images). By iteration, a realistic result emerges from a universe of possibility, and we have an intimation of godhead, albeit one that my students consistently find creepy, as the adversarial network literally draws into being a set of persons in the world who never actually were, children who are not children of god (a concept which takes on a nearly facetious register in McCarthy's *Child of God*, since, if god exists, god is also the creator of all things dull and nasty, too).

In the case of this thought experiment, how would the discriminative network do its work, since, by definition, *god is not* "it"? "Success" in proximity to god might come to look like whatever results most closely corresponded to and patterned the observable universe—the best simulation wins in a world of false starts with emergent systems and basic rules, working backwards from negative space. A busy generative network constantly told "that's not it," with enough data and time would keep monkeying with the possibilities until Hamlet, or, say, a plausible universe, hove into view. "That meek darkness be thy mirror," advises the anonymous fourteenth-century mystic author of *The Cloud of Unknowing,* a darkness, suggests editor Evelyn Underhill, that "is the 'night of the intellect' into which we are plunged when we attain to a state of consciousness which is above thought; enter on a plane of spiritual experience with which the intellect cannot deal." This, the medieval scribe says, is "the mysterious radiance of the Divine Dark, the inaccessible light wherein the Lord is said to dwell, and to which thought with all its struggles cannot attain."[40]

In *The Crossing,* the gypsy of the biplane salvage crew speaks of a God who deliberately conceals the future, such that those "who by some sorcery or by some dream might come to piece the veil that lies so darkly over all that is before them may serve by just that vision to cause that God should wrench the world from its heading and set it upon another course altogether and then where stands the sorcerer? Where the dreamer and his dream (407)?" Here is the apophatic God of the void, of the not-it, concealed behind the cloud of unknowing in the radiance of the Divine Dark (could it be dark matter?).

There's a larger point as regards apophatic theology, too: in the search for the fountainhead of creation, enormous expanses of nothingness/not-it would facilitate the infinitely shifting and scaling and emergent nature of what exists. The best simulation might only be mapped if connected by emergent nodes of the not-it, impenetrable zones of informational dark matter moving away from what exists, chased by the algorithm with the paradoxical effect of making the *what is* legible, even while paradoxically affirming that it cannot be fixed, and never closing on the end of the simulation. (Recall Vopson's hypothesis, in Chapter 4, that unexplained dark matter is in fact information, or, if you like, a revelation of the outer extent of the bounds of a simulation.)

Such a simulation posits an emergent theology alongside an emergent universe; along with the *I-am-who-am* comes the *I-am-what-is-not,* a god behind the black hole of information collapse and impossibly illegible and shifting information, the god of the next loop. In Hegelian terms, as Belfast philosopher Peter Rollins argues, the reason that atheisms produce new theisms is that thesis and antithesis do not negate one another, but, rather, produce a new synthesis (per the earlier *saying-god-does-not-exist-ipso-fact o-exists-him*). And, as Rollins might add, the common misunderstanding is that thesis and antithesis are "opposites." Even being and not being need not be opposites but complementarities; the problem, in one sense, is that our existence makes it possible to imagine the non-existence that Alicia craves in *Stella Maris*, which is absolute—not merely not to live or have lived, but never to have been. The instability of our simulation in which *god is not what you say* he/she/they/zir/his dudeness *is* shares in Gödel's problem— whatever is generated will be unprovable and incomplete within its own bounds. The statement is an antidote to the problem of "trying to *say* things that are only *showable*," in the way that an equation is a secondary artifact, at best, a photograph of reality, or perhaps, as Wittgenstein later concluded, "it is in the activity of science, whether via experiment or calculation, that all the useful information it generates exists."[41] Moreover, as Rollins puts it, "God's interaction with the world is irreducible to understanding, precisely because God's presence is a type of hyper-presence," the field from which we cannot stand outside.[42]

Back to our simulation, which describes realities that do not exist (and have no physical existence unless information has mass, which most physicists do not accept), or, at least, do not exist until they do: the nodes of the not-it, at once infinitely dense and swarming outside of the set, are drawing, counterintuitively perhaps, an adversarial universe into sharper focus ("salvaged out of a bleak sea of the incomputable" as Alicia Western says in *Stella Maris*, mere "forms turning in a nameless void" [155]). The premise of my simulation is the incomputable, ever emerging from behind the cloud of unknowing. In *Stella Maris*, Alicia Western offers an imaginative simulation of the beginnings of things, conceived at the tender age of twelve:

> One of the things I realized was that the universe had been evolving for countless billions of years in total darkness and total silence and that the way that we imagine it is not the way that it was. In the beginning always was nothing. The novae exploding silently. In total darkness. The stars, the passing comets. Everything at best of alleged being. Black fires. Like the fires of hell. Silence. Nothingness. Night. Black suns herding the planets through a universe where the concept of space was meaningless for want of any end to it. For want of any concept to stand it against. And the question once again of the nature of that reality to which there was no witness. All of this until the first living creature possessed of vision agreed to imprint the universe upon its primitive and trembling sensorium and then to touch it with color and movement and memory. It made of me an overnight solipsist and to some extent I am yet. (40)

How different we are to a "trembling sensorium"—an optic nerve with some cells to excite—or perhaps a single-celled paramecium, is a question taken up in a thought exercise posed by SFI professor David Wolpert:

Consider a single-celled, oblong paramecium, the kind that float in oceans or stagnant pools. It may seem obvious, but a paramecium—like my dog—cannot conceive of the concept of a 'question' concerning issues that have no direct impact on its behaviour. A paramecium cannot understand the possible answers we have considered for our questions concerning reality, but neither would it understand the questions themselves. More fundamentally, though, no paramecium can even conceive of the possibility of posing a question concerning physical reality.

Wolpert goes on to suggest that there is little reason to be confident of humans' ability to pose a question about physical reality, either: "Just as the notion of a question is forever beyond a paramecium, are there cognitive constructs that are necessary for understanding physical reality, but that remain unimaginable due to the limitations of our brains?" *Stella Maris* shows how decades in situ at the Santa Fe Institute and conversations with thinkers like Wolpert influenced McCarthy's thinking. In *Stella Maris* these ideas are recast and interrogated in dramatic dialogue, and sometimes not very dramatic dialogue. "We can conceive of many things even if they can never be 'known.' But among those things that we can never know is a strictly smaller subset of things that we cannot *imagine*. The issue is what we can ever perceive of that smaller set," contends Wolpert.[43] That we can know something about that which we truly can't imagine is an idea that has been around for a long time, and provides the basis for Aquinas's metaphysics, beginning with the notion that for the set of things called "god" matter and motion are required neither for being nor being understood. Even so, if godhead, which is hypernonymous (in Peter Rollins's coinage—infinitely, variously named), evades human finitude and imagination, how much more so, human description and our small grammar and symbol set? Wolpert superciliously suggests that it is rather a large coincidence that "the form of human mathematics, and of our sciences more generally, just happens to exactly coincide with the form of human language," like McCarthy, questioning what physicist Eugene Wigner famously termed "the unreasonableeffectiveness of mathematics in the natural sciences." Wolpert adds, "A cynic might also wonder whether an ant, who is only capable of formulating the 'rules of the Universe' in terms of pheromone trails, would conclude that it is a great stroke of fortune that ants happen to have the cognitive capability of doing precisely that."

Recall Davenport's archetypal researchers—patterning the use of vegetables in literary imagery, no less—who dared to ask, "quite seriously, if literature writes itself. Their theory approaches what we have all suspected, that culture is a language of images and ideas, and functions according to a syntax, with dialects and idiomatic constructions." If so, complexity theory would reveal how these emergent patterns evolve from simple rules. But we should ask as well, how is this grammar of narrative imagination imprinted in us, such that "literature writes itself"? (McCarthy waves a flipper at the unconscious here.) Have we improved much upon Jung's suggestion that it resides in the mystery of the unconscious? And does this world of forms confirm the rather limited universe of narrative patterns, pointing in turn to our limited, paramecium-plus perception? Can it any way conduct us to the realm of things that cannot be imagined?

In *All the Pretty Horses*, McCarthy's most obsessive exploration of Platonism, John Grady Cole's dream of horses unfolds in a Faulknerian sentence of 211 words, ending with a sort of *gloria* for the world of forms:

> and they flowed and changed and ran and their manes and tails blew off of them like spume and there was nothing else at all in that high world and they moved all of them in a resonance that was like a music among them and they were none of them afraid horse nor colt nor mare and they ran in that resonance which is the world itself and which cannot be spoken but only praised. (161)

The passage presents McCarthy writing in his most positive aspect. His brother Dennis dwells more comfortably in this territory with his first novel, *The Gospel According to Billy the Kid* (2021), which offers an exploration of spiritual development through an unlikely vessel (yes, yet another McCarthy Kid). Where Cormac's beliefs hew to agnosticism if not atheism, Dennis's spiritual practice and writings dwell in the deserts of Richard Rohr-esque Catholic mysticism in a syncretic matrix where Christianity intersects with other wisdom traditions, including Buddhism. Dennis and his wife are oblates at Christ in the Desert Monastery, and while working on the novel, he experienced a revelation—some visitation of spiritual and fictional certitude—that Billy the Kid had once been among the monastics in New Mexico. Or would be for purposes of the story. Immediately after gut-churning scenes of destruction and torture in the novel based on actual events, in one of the book's most striking dialogues, one of the monks shares a version of the Navajo beauty prayer with the shaken young gunslinger, the novel's namesake: "With beauty before me I walk. With beauty behind me I walk. With beauty around me I walk. . . . It is finished in beauty."[44]

In Cormac's work we find the paradox of beautiful destruction and in Dennis's we see the paradox of a faith pinned between conflicting indications of divine presence, dramatized in the mystery of a gunslinging kid's salvation. Cormac's musings rarely dwell for long on that which can only be praised, more often tipping sharply toward sublime terror and solipsism instead. Would Alicia's universe emerge in a simulation derived from apophatic theology in which a universe is simulated not by conjuring upon what is, but rather by what it is not via the rule that the universe is not what you say it is? Meaninglessly large spaces and multiverses, the impossibility of understanding nothing, the terrifying 0 to the digital 1 of our felt existence? Eventually something emerges in relief by a series of increasingly accurate approximations of negative space that race endlessly ahead of our small perception and intelligence, indeed, "for want of any concept to stand against it," in Alicia's phrase. As soon as one creates a system capable of self-reference, it will be incomplete, because if it is accurate, it won't be able to express all truths about itself. Also, in a simulated universe, Turing's halting problem applies, as our simulation will never know when it can stop, even if it could somehow acquire all the information in the available universe, and sufficient energy for running the simulation. The idea of the computational power to run such a simulation might not be purely the stuff of science fiction if quantum computing continues to develop—"If and when a quantum computer is built using 275 qubits," IBM has claimed, "it could represent more computational states than there are atoms in the known universe."[45] IBM has promised 1000 qubits by 2023 and by the end of 2021, and Google wants to produce a million-qubit computer within ten years, naturally, which would hasten the arrival of vaunted artificial general intelligence.[46] Some argue that the conditions required for commercially viable quantum computing that will make present-day computers "look like toys" will never be feasible, and that quantum computing is a media-hyped technological will-o'-the-wisp in a class with practical superconductors and cold fusion, which always seem just out of reach.[47]

Regardless, it seems to me that even with more data and better measurement and no matter how much energy, the god in the machine would slip away once again as we came to the edges of creation and arrived at a coastline paradox, that standard reference of complexity science—an endlessly receding shoreline of what god is *not*, infinitely divisible, which is also part of what god (or reality) *is*.[48]

By now some readers will have noticed that the proposed simulation describes one vision of a simulated world theory. Simulation theory was once largely a laughingstock, taken seriously by a select company of philosophers and the occasional colorful Musk-ovite, yet a growing number of theorists acknowledge that it is at least possible that we are merely qualia in service of a complex system, or a greater consciousness generated by simulation, a notion that McCarthy approaches in the 2022 novels by probing where the unconscious arises from the complex and unknowable (more on this momentarily). His fellow SFI wayfarer David Wolpert urges that the question of whether we exist in a simulation is "rather trivial" as he is more interested in whether we "might be able to establish the existence of what we can never conceive of through observation, simulation, theory or some other process" and so be freed from our very limited conception of physical reality. The duology of McCarthy's final works dramatizes through the persistent harangues of the Kid that the unconscious might be the porter to this kingdom beyond the cloud of unknowing—even though we swat it away, deriding its "antics" and "primitive understanding." Whereas McCarthy posits that the unconscious is a machine for operating an animal, Fouad Kahn somewhat fatuously suggests in *Scientific American* that "the simplest explanation for the existence of consciousness is that it is an experience being created, by our bodies, but not for us." Khan was poking fun at the vogue for simulation theory in *Scientific American* ("Confirmed! We Live in a Simulation"), concluding, superciliously, that "We must never doubt Elon Musk again."[49]

"If we are living in a simulation," philosopher Nick Bostrom wrote in his seminal 2003 paper on the topic,

> then the cosmos that we are observing is just a tiny piece of the totality of physical existence. The physics in the universe where the computer is situated that is running the simulation may or may not resemble the physics of the world that we observe. While the world we see is in some sense 'real,' it is not located at the fundamental level of reality.[50]

Naming recent adherents, including Neil deGrasse Tyson, Khan also cites skeptics such as physicist Frank Wilczek (who "has argued that there's too much wasted complexity in our universe for it to be simulated") and physicist Sabine Hossenfelder, who points out that the hypothesis cannot be falsified.

The tongue-in-cheek article attempts to race to the edges of the simulation's computing architecture. "All computing hardware leaves an artifact of its existence within the world of the simulation it is running. This artifact is the processor speed," he explains, and concludes that the speed of light coincides with the processor speed, hence stands as the artifact of our simulation. In this way the theory provides an answer to one of the inferred questions McCarthy asks in the Kekulé essays, though he poses it more directly of the unconscious—where did consciousness come from? To provide the simulation creator with the qualia of our experience, Khan replies, with a wink.

The problem with simulation theory is, in a sense, Wittgenstein's anticipated problem with simulation theory, and McCarthy's problem with language: it is merely secondary description of what appears to exist, not just metaphysically unreliable, but calling into question any metaphysic. A durable type of human venality renders us susceptible to thinking that better description could enable us to recreate the mind of god, which is not what you say it is. One could say that it is simply the latest version of Descartes's evil demon problem, later the brain-in-a-vat problem, now the simulation problem. It does not follow from our ability to create another consciousness that we are the product of such a consciousness or imprisoned by it. Looking around the world today, most people have their memory and consciousness extended by an object generally pocketed at buttocks-level (a phone), which creates both artificial intelligence and stupidity.

Moreover, as David Krakauer suggests, the vaunted singularity itself is an emergent, ancient phenomenon. "There's been a lot of talk about the singularity, a consequence of miniaturization in silicon chips, that at some point the memory capacity and the processing speed will be sufficient for human beings or some other organism to create a simulacrum of themselves in a computer," he submits.

> But humans already did that in Sumeria in the Third Millennium BC when they inscribed their thoughts in clay in cuneiform using a reed. There was some version of you, some dilute version of you in a physical artifact. It wasn't all of you, but there never will be all of you, because even a downloaded version of your mind would lose your physical body. So I call that the first singularity, and it wasn't a singularity that came from the silicon chip. It came from the silicate chip, the clay chip, and history has been in some sense a whole series of inventions that have allowed us to create or deposit versions of ourselves. Think about novels, think about the *Epic of Gilgamesh*. It is a distillation of emotional, moral, ethical ideas that society was obsessing over and for people to have access to those ideas in distilled form must have been revolutionary.[51]

Yet simulation theory is something that complexity science must take seriously, and McCarthy flirts with it in his final duology. In *The Passenger* Bobby Western recollects Alicia saying, "There will be nothing that cannot be simulated. And this will be the final abridgment of privilege. This is the world to come. Not some other. The only alternate is the surprise in those antic shapes burned into the concrete (382)." So Alicia bounds the ineluctable movement of earthly complex systems toward simulated reality with only the shadows of Hiroshima—the movement toward converting earth energy and systems into data will be stopped by an unanticipated, human-caused extinction event. For now, we are coursing along the edge of the circuit. What makes complexity fascinating is precisely what happens at the edge of chaos. Per Dowd,

> Structures that improbably weave complex yet fleeting intricacies of order amid the irreversible flow of thermodynamic entropy, of which living organisms are but one example, were described by Nobel laureate chemist Ilya Prigogine as dissipative structures. Prigogine's work laid some of the groundwork for contemporary complexity science, in his demonstrations that the flow of energy from ordered to disordered states can give rise to self-organizing structures that thrive on the edge of chaos. Such structures tap into the thermodynamic traffic in their spontaneous emergence and self-organization, utilizing the flux of energy passing through them to form islands of complex order that temporarily and locally stand against the universal drift to entropy inscribed in the second law of thermodynamics.[52]

Emphasis on "temporary": the second law ultimately prevails, so that all we know of negentropy—the opposite of entropy—in this world are these sparks on the edge of disorder that briefly become more organized, including, for a time, the self-organizing and scalable properties of complex systems derived from simple rules that we describe, simply, as *living*. In our mind-exploding simulated universe, we have unleashed a hunter on a never-ending stalk in search of itself by way of its creator. It would seem to befit the vanity of human wishes, this Melvillian pursuit of the whale, when, as McCarthy's screenplay *Whales and Men* intimates, we are inadequate to "deal with the whaleness of the whale":

> There are times when I see the whale very differently. When I get a fleeting vision of the pure platonic whale and I have a sense that what we are after, the whaleness of the whale, does exist as an idea, but an idea with which we are inadequate to deal. I think that's why we kill them. I think their value as produce is secondary to their sacrificial value. It's our inadequacy to the overwhelming fact of the whale's existence that will doom the whale. I think Melville is right. It's like killing God. That whale's existence is the whale I cant deal with. That it is the whale who strings the world together on the vectors of his breath and I wont know in this world where that whale will have gone when the last whale is slaughtered and hauled from the sea.
>
> But that there can be no whale seems a sort of monstrous paradox. As if we were to say that the universe, if it should contract again into the singularity from which it sprang, would then no longer exist. According to whom? Nonbeing requires a witness the same as being does and if the whale should come to no longer exist, still in some way I think I would disbelieve in our competence to say that that was so. (He smiles) At other times of course I see only the empty silent seas through which he has passed forever. (25–26)

Whales and Men channels indirectly conversations with cetologist Roger Payne, McCarthy's friend for decades, in a screenplay that speculates about the possibility of a continuous narrative of creation, housed in the brains of and spoken in the oral tradition of whales, who, after all, have been communicating, with larger brains, in their environment, for a very long time as a species, and whose phrases might be rendered decipherable by AI, according to recent articles. The finitude and apparently temporally finite bounds of human imagination quail alike before infinitude and nonbeing, the paradox of bearing witness to them, and whether they possess a reality independent of observation. Reflecting on expansion and contraction—the spiraling gyres of the physical universe—in an unpublished fragment of *Whales and Men*, McCarthy writes, "If this looping process is the way it works then there's no need for teleology." The arrow of time, teleology, a syntax that produces meaning—these are elements by which we shore seawalls against empty silent seas, which appear in the unmade world of *The Road*, and the Gulf shoreline which Bobby Western walks in *The Passenger*. Were the arrow of time reversed, were the world unmade, were the whalebones scattered on the desolate shore (as they are in the final chapters of *The Road*), were there no witness, could there still be meaning without "purpose"? Why should animation occur in the unlikely contact zone between the inert and the entropic, between consciousness and the unconscious (a mystery Annaka Harris probes in *Conscious: A Brief Guide to the Fundamental Mystery of the Mind* (2019))?

In *The Passenger*, the Thalidomide Kid, who in my reading is the allegorical voice of the unconscious, is depicted in the text, ambiguously, as a possible hallucination—but so are we. As the Kid says to the consciousness he inhabits,

> *We got some herky jerky images of dudes and dudesses but they got no name. They used to have names but they dont any more. The last witness who could have put a name to the faces is boxed up in the ground alongside them and if not nameless as well will soon be so. So. Who are they? The fact that they once walked about in the nomenclative mode is small comfort. Small comfort to whom? Well shit. You just throw up your hands.* (191)

Put differently, the naming of things including people might not give them lasting meaning,[53] but rather nomenclature and emerging taxonomies simply acknowledge the sparks at the edge of chaos, the shared activity of bearing witness to miracles of complexity at a moment in time. In the post-apocalyptic world of Cormac McCarthy's *The Road*, a boy and his father encounter a murderous man, a destroyer "who has made of the world a lie every word" (75). Yet the novel also suggests that ritual, litany, rite, "making up a cheerful song" can call goodness into being, suggesting that words can call certain realities into being. Is this a contradiction, and where does the power come from? Einstein's equations did not generate reality so much as afford insights into its underpinnings.

Likewise, when James Cornman and Keith Lehrer dreamed up their "braino cap" hypothetical in 1968, in which a "braino machine" could produce hallucinations according to the operator's desires, it forecast the current era of *Meta* experience and simulation, and soon became a staple of science fiction film.[54] They concluded that, while we might sometimes be able to tell when we are hallucinating, it would not be possible to know definitively that our experience writ large is not a hallucination. Certainly, this thought experiment seems less abstract in the current of age of virtual and augmented reality, synthetic media, and screen-mediated reality, and when, for example, artificial intelligence troubles patent law by creating its own born-digital intellectual property synthesized from human discoveries.[55] To whom shall those ideas be attributed?

The Santa Fe Institute and Krakauer are pondering these questions and the possibility that the next great unifying theory of the future might come not from a mathematical proof but in the form of a computer program, and that it will not necessarily be a product of human intelligence. "Everything worthwhile is an effort and we celebrate it but that might go away when the things overcoming the obstacles are no longer human beings. What happens to the motive force driving culture forward when examples of success have disappeared?" Krakauer asks. "I think the cultural implications of surrendering very difficult problems might be very significant and we haven't really thought it through."[56]

Already, columnist Farhad Manjoo submits, "The mingling of physical and digital reality has . . . thrown society into an epistemological crisis—a situation where different people believe different versions of reality based on the digital communities in which they congregate."[57] Moreover, we are fast approaching a point where natural language generation algorithms (NLGs) are making AI-generated prose indistinguishable from the human-generated variety, raising the possibility (or likelihood) that the next Cormac McCarthy could be Cormat McBothy.[58] If this troubles you, your only consolation might be "that the future distant relatives of modern kitchen robots will not turn out to be much smarter than ourselves," as Karol Jałochowski—physicist, journalist, documentarian, and friend of SFI—wrote in 2008.[59]

Whether simulation leads to a strong or weak AI, per John Searle, is irrelevant. If my theory that the Thalidomide Kid of the duology directly represents the unconscious, alternate modes of interpretation arise: he could be the hiccup or artifact in the simulation, some proof of our non-machineness, or something else altogether. You throw up your hands. The text of *Stella Maris* is coy about whence he came:[60]

> The Kid.
> Oh. Yes. By whom was he sent then?
> I dont know. He's no more mysterious than the deeper questions about any other reality. Or mathematics. For that matter. Forms turning in a nameless void. Salvaged out of a bleak sea of the incomputable. Time's up. (155)

The time is up for this study, the time is nigh for this experiment—though it should yield only a beautiful failure, may it be so, and the beginning of something with no probable conclusion.

Notes

1   "#20" from *Hexagrams: A Poet's Journey Through the I Ching* (Winston-Salem, NC: WFU Press, 2023), 54. Reprinted by permission of the author and Wake Forest Press.

2   I am grateful to Dennis McCarthy for making the early draft available to me.

3   Bryan Giemza, "Guy Davenport," in *The South Carolina Encyclopedia* (University of South Carolina, Institute for Southern Studies, 2016), https://www.scencyclopedia.org/sce/entries/davenport-guy/.

4   John J. Sullivan, "Guy Davenport, The Art of Fiction No. 174," *The Paris Review* 163 (Fall 2022): n.p.

5   I benefited from Davenport's generous critical eye, too, as he was once a peer reviewer for an unconventional journal article I wrote, as I would learn much later. "This essay is scurrilous, irreverent, and bizarrely creative," he wrote. "I think we should publish it."

6   Letter from Cormac McCarthy to Guy Davenport postmarked May 12, 1978, Guy Davenport Papers, Letters from Cormac McCarthy, 1968–1989 and undated, Container 133.2, Manuscript Collection MS-4979, Harry Ransom Center, The University of Texas at Austin.

7   Letter postmarked November 10, 1984, Guy Davenport Papers. Presumably McCarthy had an advanced reading copy or other draft. He knew both Feynman and Gell-Mann and their complex relationship.

8   N.d., Guy Davenport Papers.

9   Letters postmarked June, 1981, July 28, 1981, October 28, 1981.

10 Letters postmarked March 13, 1985 and February 25, 1986, Guy Davenport Papers.

11 For Beckett's influence on McCarthy see Richard Rankin Russell, "'On': Reading Cormac McCarthy's *The Road* through Beckett's *Waiting for Godot* and *Ill Seen Ill Said*, *English Studies* 101, no. 5 (January 2020), https://doi.org/10.1080/0013838X.2019.1695178.

12 Letter postmarked July 28, 1981, Guy Davenport Papers.

13 N.d. [1981], Guy Davenport Papers.

14 Guy Davenport, "The Critic as Artist," in *Every Force Evolves a Form* (San Francisco, CA: North Point Press, 1987), 107. The study Davenport references can still be found among online booksellers. Ralf Norrman and Jon Haarberg, *Nature and Language: Semiotic Study of Cucurbits in Literature* (London: Routledge, 1980).

15 For a fascinating exploration, see Carmine Starnino, "Poetry and Digital Personhood," *The New Criterion*, April 2022, https://newcriterion.com/issues/2022/4/poetry-digital-personhood.

16 Davenport, "The Critic as Artist," 100.

17 Cormac McCarthy to J. Howard Woolmer, Undated [1986], Woolmer Collection of Cormac McCarthy, Southwestern Writers Collection, Texas State University-San Marcos, Box 1, Folder 6.

18 "Tracing Languages Back to Their Earliest Common Ancestor through Sound Shifts," Santa Fe Institute, February 19, 2015, https://www.santafe.edu/news-center/news/bhattacharya-pagel-sound-shifts.

19 The familiar "Indo-European" branch is simply a convenient catchall term that neglects the African origins of humankind. Given the small numbers of early kin groups and societies, it seems most likely that what happened might resemble spontaneous and potentially even convergent evolution based on the simplest set of vocalizations arising in multiple locales simultaneously. SFI investigators acknowledge that applying nonlinear dynamics and evolutionary biology to language is an uncertain business; in the best case, it might yield a crude look at the whole.

20 Murray Gell-Mann and Merritt Ruhlen, "The Origin and Evolution of Word Order," *PNAS* 108 no. 42 (October 10, 2011), doi: https://doi.org/10.1073/pnas.1113716108.

21 "Projects: The Origins, Evolution, and Diversity of Human Languages," The Santa Fe Institute, https://www.santafe.edu/research/projects/the-origins-evolution-and-diversity-of-human-langu.

22 Carl Zimmer, "Nonfiction: Nabokov Theory on Butterfly Evolution is Vindicated," *New York Times*, January 25, 2011, https://www.nytimes.com/2011/02/01/science/01butterfly.html.

23 Cormac McCarthy, "The Kekulé Problem: Where Did Language Come From?" *Nautilus* 47 (April 20, 2017), http://nautil.us/issue/47/consciousness/the-kekul-problem. Similarly, Alicia pints to the strange origins of "realize" in *Stella Maris*: "Odd locution. Literally of course it means to make real. What if I just told you some monstrous lie?" (157)

24 Davenport, "The Critic as Artist," 111.

25 "One Reporter's Account of SFI's 'Genius and Madness' Event in August," Santa Fe Institute, November 9, 2015, https://www.santafe.edu/news-center/news/genius-and-madness-sz-trans-english. Translated and reprinted with permission from Ulrike Duhm's reporting on "Genius and Madness" in the German daily *Sueddeutsche Zeitung*.

26 Alexandra Alter, "Sixteen Years After *The Road*, Cormac McCarthy Is Publishing Two New Novels," *New York Times*, March 8, 2022, https://www.nytimes.com/2022/03/08/books/cormac-mccarthy-new-novels.html.

27 Jonathan Miles, "Double Punch," *Garden & Gun*, October/November 2022, 46.

28 John Jurgensen, "Hollywood's Favorite Cowboy," *Wall Street Journal*, November 13, 2009, https://www.proquest.com/newspapers/hollywoods-favorite-cowboy/docview/399070032/se-2?accountid=7098.

29 Scientists continue to refine methods for controlling enantiomers at the molecular level. Cf., Fritz Haber Institute of the Max Planck Society, "Controlling Chemical Mirror Images," *Phys.org*, April 22, 2022, https://phys.org/news/2022-04-chemical-mirror-images.html.

30 Paul Krugman, "The Strange Alliance of Crypto and MAGA Believers," *New York Times,* January 10, 2022, https://www.nytimes.com/2022/01/10/opinion/crypto-cryptocurrency-money-conspiracy.html.

31 Cormac McCarthy, "Connecting Science and Art," *Science Friday* interview by Ira Flatow, National Public Radio, April 8, 2011, https://www.npr.org/2011/04/08/135241869/connecting-science-and-art.

32 Alter, "Sixteen Years After *The Road.*"

33 Jurgensen, "Hollywood's Favorite Cowboy."

34 Alter, "Sixteen Years After *The Road.*"

35 Ultimately, I have concluded that Alicia is an amalgamation of several living women McCarthy is acquainted with, including, primarily in her brilliance, the physicist Lisa Randall.

36 Willard Van Orman Quine, "On What There Is," in *Western Philosophy: An Anthology*, ed. John G. Cottingham (Hoboken, NJ: WileyBlackwell, 2021), 144.

37 The caves continue to inspire art and new commentary. For example, in 2020 Tim Heidecker wrote the lyrics to a song based on the Chauvet caves in which he imagined women painting on the wall, asking, "I wonder if they ever dreamt of us at all." Songwriter and singer Laura Mering then set the words to music, and the resulting song, "Oh How We Drift Away," is a transporting meditation on how we continue to paint over the cave walls of our deep ancestral past.

   Tom Breihan, "Weyes Blood & Tim Heidecker—'Oh How We Drift Away,'" *Stereogum*, September 22, 2020, https://www.stereogum.com/2098947/weyes -blood-tim-heidecker-oh-how-we-drift-away/music/.

38 McCarthy, "Connecting Science and Art."

39 *Digital Transformation: Interview with David Krakauer*, directed by Manuel Stagars (August 2017), streaming documentary video, https://digitaltransfor mation-film.com/david-krakauer-santa-fe-institute/.

40 The proper title of the second edition edited by Evelyn Underhill, directly from the British Library original manuscript, is *A Book of Contemplation the Which Is Called the Cloud of Unknowing, in the Which a Soul Is Oned with God* (London: John M. Watkins, 1922), 30.

41 Timothy Andersen, "Quantum Wittgenstein," *Aeon*, May 12, 2022, https:// aeon.co/essays/how-wittgenstein-might-solve-both-philosophy-and-quantum -physics.

42 Peter Rollins, *How (Not) to Speak of God* (Paraclete, 2006), 23.

43 Consider how the anonymous mystic who penned *The Cloud of Unknowing* attempted to frame knowledge of unknowable in a fashion not dissimilar to Wolpert: "And ween not, for I call it a darkness or a cloud, that it be any cloud congealed of the humours that flee in the air, nor yet any darkness such as is in thine house on nights when the candle is out. For such a darkness and such a cloud mayest thou imagine with curiosity of wit, for to bear before thine eyes in the lightest day of summer: and also contrariwise in the darkest night of winter, thou mayest imagine a clear shining light. Let be such falsehood. I mean not thus. For when I say darkness, I mean a lacking of knowing: as all that thing that thou knowest not, or else that thou hast forgotten, it is dark to thee; for thou seest it not with thy ghostly eye. And for this reason it is not called a cloud of the air, but a cloud of unknowing, that is betwixt thee and thy God." Underhill, *A Book of Contemplation*, 72.

44   Dennis McCarthy, *The Gospel According to Billy the Kid: A Novel* (Albuquerque, NM: University of New Mexico Press), 132.

45   "Exploring quantum computing use cases for manufacturing," https:// www.ibm.com/thought-leadership/institute-business-value/report/quantum -manufacturing.

46   Adrian Cho, "IBM Promises 1000-Qubit Quantum Computer—a Milestone— by 2023," *Science,* September 15, 2020, https://www.science.org/content/article /ibm-promises-1000-qubit-quantum-computer-milestone-2023.

47   Cade Metz, "'Quantum Internet' Inches Closer with Advance in Data Teleportation," *New York Times,* May 25, 2022, https://www.nytimes.com /2022/05/25/technology/quantum-internet-teleportation.html; https://www .polityka.pl/tygodnikpolityka/nauka/1506974,1,kryptografia-kwantowa-co -to-takiego.read. But see Subhash Kak, "A Quantum Computing Future Is Unlikely, due to Random Hardware Errors," *The Conversation*, December 3, 2019, https://theconversation.com/a-quantum-computing-future-is-unlikely -due-to-random-hardware-errors-126503.

48   The epigraphs for Davenport's titular essay, "Every Force Evolves a Form," are from (1) the non-canonical Gospel of Thomas (*Jesus Said: Split a Stick. I Will Be Inside*) and (2) Emily Dickinson (*Split the Lark, and You'll Find the Music,/ Bulb After Bulb, in Silver Rolled*), 151.

49   Fouad Khan, "Confirmed! We Live in a Simulation," *Scientific American*, April 1, 2021, https://www.scientificamerican.com/article/confirmed-we-live-in-a -simulation/.

50   Nick Bostrom, "Are You Living in a Computer Simulation?" *Philosophical Quarterly* 53, no. 211 (2003): 243–55, https://www.simulation-argument.com/ simulation.pdf.

51   *Digital Transformation: Interview with David Krakauer.*

52   Ciarán Dowd, "The Santa Fe Institute," in *Cormac McCarthy in Context,* ed. Steven Frye (Cambridge: Cambridge University Press, 2020), 33–44, doi: https://doi.org/10.1017/9781108772297.005.

53   Here I am contra Quine on the ontological problem (that naming what does not exist brings it into existence): "Names are . . . altogether immaterial to the ontological issue, for [they] can be converted into descriptions," and by the use of what logicians refer to as bound variables, "descriptions can be eliminated." Quine, "On What There Is."

54   Cf. James W. Cornman, George Sotiros Pappas, and Keith Lehrer, *Philosophical Problems and Arguments: An Introduction* (Indianapolis, IN: Hackett, 1992), chapter 2.

55   Alexandra George and Toby Walsh, "Artificial Intelligence Is Breaking Patent Law," *Nature*, May 24, 2022, https://www.nature.com/articles/d41586-022 -01391-x.

56   *Digital Transformation: Interview with David Krakauer.*

57 Farhad Manjoo, "We Might Be in a Simulation. How Much Should That Worry Us?" *New York Times,* January 26, 2022, https://www.nytimes.com /2022/01/26/opinion/virtual-reality-simulation.html.

58 Starnino, "Poetry and Digital Personhood." There is strong commercial incentive for this technology as producers seek to avoid paying licensing fees to creatives. For example, AI-generated soundtracks, jingles, and scores are already being produced, obviating the (expensive) need for composers, musicians, studio time, and so on.

59 Karol Jałochowski, "A Lover with a Switch," *Polityka,* July 26, 2008, https://www.polityka.pl/tygodnikpolityka/kultura/262177,1,kochanek -zwylacznikiem.read.

60 It is also coy about the parentage of the Western "siblings." The text spreads ambiguity around the details of Alicia's birth and parentage, with various inconsistencies and insinuations. For example, when Alicia relates the details of her visit to her putative father's "pretty dishy" (76) first wife in California— after her father's death—she insistently refers to her deceased (putatively half-brother) by that first union: "He was my brother," and again, "My brother's name was Aaron" (76). Divorced California wife says curiously of Alicia, "You turned out all right," and Alicia says also of California wife, "she'd heard rumors about me and she figured I'd show up sooner or later" (76). The same psychiatric dialogue hints that her father was a "philanderer" (76) who was away from the family for long periods, and relates what Alicia says of her parents' eventual estrangement and her mother: "I think they loved each other. It just became more and more difficult. She looks nervous. She's smoking faster. Of course this could be a false tell. She's a devious little bitch" (77). The term *false tell* relates to McCarthy's interest in gambling, denoting the strategy of playing a losing card to deceptive purpose. Other known facts about Alicia's Tennessee mother: she was exceptionally bright, worked as a calutron girl in Oak Ridge amid a steady circulation of young men staying in dormitories and barracks, accepted Mr. Western's proposed date on his second approach via a slipped note by writing down her dormitory's phone number, and was reportedly the most beautiful woman in the state at that time, in a region where Oak Ridge offered the best prospects for decent jobs and marriage, at a time in a male-dominated culture when (McCarthy makes clear) very bright women had few economic options. What follows are not the most obvious interpretations, and indeed might not satisfy Occam's razor, but McCarthy seems to open the door to these at least *plausible* possibilities: 1) California mother, also Jewish, is Alicia's real mother, and Aaron, who died of polio at four years of age, was Alicia's biological brother, and she grew up, unwittingly, in a blended family, with Bobby as her biological half-brother (contra Alicia's assertion that her mother would not have tolerated knowledgeof her father's previous marriage); or 2) Alicia and Bobby are biologically unrelated owing jointly to "devious" Tennessee mother's infidelity and father's "philandering," papered over by a marriage of convenience to conceal a blended family. Admittedly, neither of these interpretations is the most obvious reading, but with deference to time, motivation, and opportunity, we might wonder why

Alicia unambiguously calls Aaron her brother and why McCarthy raises a poker tell from his unreliable narrator—not to mention why she seems to have no innate revulsion to her attraction to her putative "brother." If this reading seems strained, it would nevertheless achieve two effects for the novel: first, it vindicates Alicia in a certain way, and second, it puts the tale on a footing of high tragedy if Alicia's singular love for her "brother" might have been permissible after all.

ACKNOWLEDGMENTS

Work is prayer and I've been praying these verses for a decade, worrying the pages like rosary beads. Yes, books go like life that way. Like all my endeavors in this chapter of life, this one would not have happened without my wife, Kristi, or my children, John Paul and Vera Rose.

Among friends and colleagues, particular thanks are due to Dennis McCarthy, a brother of the heart. Without his generosity and encouragement this book could not have happened, or at least, not as it happened. His friendship has been a signal blessing to our family since we moved southwest. Thanks, bud.

I am very grateful to the McCarthy family writ large, including Judy McCarthy, Anne De Lisle, Barbara (Bobbie) McCooe, and Cormac McCarthy, for sharing their time and memories. This material will find its place in the literary record in the fullness of time; meanwhile, these pages are informed by the real human understanding they brought to them.

At the Santa Fe Institute, immense gratitude to David Krakauer and Tim Taylor, for cracking conversation, gracious hospitality, and the sort of inspiration needful for projects of endurance and delight.

For underwriting research in various capacities, I am grateful to the Texas Tech Honors College, the Texas Tech Humanities Center, the Wittliff Collection, and James E. Sowell.

Many friends and colleagues and members of the Cormac McCarthy Society have strengthened this book by reading chapters or sparking fresh ideas. These include Scott Yarbrough, Stacey Peebles, Rick Wallach, Allen Josephs, Richard Russell, Lydia Cooper, Chris Newton, Bland Simpson, Katherine Larson, Christopher Witmore, Andy Wilkinson, Katharine Hayhoe, and Steven Frye. Here again, Dennis McCarthy's advice was indispensable to taming one particularly unruly chapter. The book benefited from the mysterious horsemen of the Infinite Bass Triangle, including Jessica Martell, Zackary Vernon, and Philip Keel Geheberer, passing together, in True Grit fashion, under the penumbra of Covid through the windswept plains of the Irish night. The chirality chapter was enhanced by conversations with Sarah Huber, Joseph Vuletich, and Michael Crews. Lewis I. Held schooled me anew in the wonders of chirality and twinning, and I count it a great serendipity to learn from a colleague whose work is field-defining. I wish to remember M. Thomas Inge, who believed in me—believed in *this*—all along.

Mary Jane Salyers shared generously of her time and memories, taking us to the authentic deep currents of colonial Appalachian experience. My thanks to her and to Joy and ecumenical friends at Watts Street Baptist Church in Durham, NC. Donald Beagle, loyal friend and Renaissance man, kindly allowed the reprint of one of his forthcoming poems on these pages. Warm thanks to him and to Celeste Holcomb at Wake Forest Press for those words.

Portions of the chirality chapter were originally published in the *European Journal of American Studies*. I am grateful to the editors for insisting on open access while providing rigorous editing. The article first appeared in 2017; it is gratifying to see chirality referenced by name in the duology, evidence, perhaps, that even a blind squirrel finds a nut once in a while.

Thanks are owed as well to the many students who through the years have taken the challenging journey into McCarthy's writing. For nearly two decades seniors and I have ridden together; even as the world at times seems to grow consecutively dark and *Road*-like, they meet it with dignity and courage. My teachings are scattered to the winds while their teachings are concentrated in this work in ways small and large. And they ride on. So I would like to especially thank Jackson Kulick, whose readings of McCarthy's works, and whose fine mind for math and science, helped me to understand so many elements in this book. Likewise, Gautam Baklikwal, who brought me up to speed on not just generative adversarial networks but the prospects of programming something like a moral code into artificial intelligence. Katie Salzmann, the archivist of Cormac's papers in the Wittliff Collection, served as an inspired and kind guide both to students and to me as this project unfolded.

Finally, large thanks are due to my editor at Bloomsbury, Amy Martin, and her team, including Hali Han. Their professionalism, kind curiosity, and respect for authors set a rare standard and brought genuine pleasure to the production of this book.

[About the Cover]

The photograph of Cormac McCarthy is from the 2010 Antarctic voyage that resulted in director Scott Cohen's debut film, *Red Knot*. The cast and crew of sixteen that sailed from Buenos Aires with Cohen included McCarthy's longtime friend, whale biology and song specialist Roger Payne, who appears in the 2014 film.

The otherworldly backdrop of red-tinted snow is eye-catching—and fitting for this writer of apocalypse, also linked to climate change. Sometimes referred to as *sang de glacier* or watermelon snow, it comes from microscopic

Chlamydomonas nivalis algae, part of a snowbound algal microbiome that blooms after a period of warmth and melting. Because the snowpack reflects less light when it is tinted, it decreases albedo—the ice's ability to return the sun's energy to the atmosphere—accelerating a warming loop and prompting some scientists to incorporate the changing hues into climate warming models.

The photograph was taken by Todd Murphy and is reproduced here by permission of Liane Murphy. The author thanks her, the Artists Rights Society, and John Poch, for bringing the image to his attention in the first place.

BIBLIOGRAPHY OF CORMAC MCCARTHY WORKS REFERENCED

For quick reference, in-text page number citations are keyed to these editions.

McCarthy, Cormac. *All the Pretty Horses*. Vintage International, 1993, orig. 1992.

McCarthy, Cormac. *Blood Meridian: Or, the Evening Redness in the West*. Vintage International, 1992, orig. 1985.

McCarthy, Cormac. *Child of God*. Vintage International, 1993, orig. 1974.

McCarthy, Cormac. *Cities of the Plain*. Vintage International, 1999, orig. 1998.

McCarthy, Cormac. "Cormac McCarthy Returns to the Kekulé Problem: Answers to Questions and Questions that Cannot be Answered." *Nautilus*, November 30, 2017, http://nautil.us/issue/54/the-unspoken/.

McCarthy, Cormac. *The Counselor: A Screenplay*. Vintage International, 2013.

McCarthy, Cormac. *The Crossing*. Vintage International, 1995, orig. 1994.

McCarthy, Cormac. "The Kekulé Problem: Where Did Language Come From?" *Nautilus*, March–April 2017, http://nautil.us/issue/47/consciousness/the-kekul -problem.

McCarthy, Cormac. "The Kekulé Problem" *Nautilus*, November 27 2017, https:// nautil.us/cormac-mccarthy-returns-to-the-kekul-problem-236896/.

McCarthy, Cormac. *No Country for Old Men*. Alfred A. Knopf, 2005.

McCarthy, Cormac. *The Orchard Keeper*. Vintage International, 1993, orig. 1965.

McCarthy, Cormac. *Outer Dark*. Vintage International, 1993, orig. 1968.

McCarthy, Cormac. *The Passenger*. Knopf, 2022.

McCarthy, Cormac. *The Road*. Alfred A. Knopf, 2006. Note: pagination is from the first edition. Subsequent editions vary.

McCarthy, Cormac. *Stella Maris*. Knopf, 2022.

McCarthy, Cormac. *The Sunset Limited: A Novel in Dramatic Form*. Vintage International, 2006.

McCarthy, Cormac. *Suttree*. Vintage International, 1992, orig. 1979.

McCarthy, Cormac. *Whales and Men*. Cormac McCarthy Collection, The Wittliff Collections, Texas State University, 91.97.5, n.d. unless otherwise specified.

INDEX

Printed in the USA
CPSIA information can be obtained
at www.ICGtesting.com
LVHW011631241123
764754LV00005B/87